ADVANCED BUILDING TECHNOLOGIES FOR SUSTAINABILITY

ASIF SYED

WILEY

John Wiley & Sons, Inc.

This book is printed on acid-free paper. ∞
Copyright © 2012 by John Wiley & Sons, Inc. All rights reserved

Published by John Wiley & Sons, Inc., Hoboken, New Jersey
Published simultaneously in Canada

No part of this publication may be reproduced, stored in a retrieval system, or transmitted in any form or by any means, electronic, mechanical, photocopying, recording, scanning, or otherwise, except as permitted under Section 107 or 108 of the 1976 United States Copyright Act, without either the prior written permission of the Publisher, or authorization through payment of the appropriate per-copy fee to the Copyright Clearance Center, 222 Rosewood Drive, Danvers, MA 01923, (978) 750-8400, fax (978) 646-8600, or on the web at www.copyright.com. Requests to the Publisher for permission should be addressed to the Permissions Department, John Wiley & Sons, Inc., 111 River Street, Hoboken, NJ 07030, (201) 748-6011, fax (201) 748-6008, or online at www.wiley.com/go/permissions.

Limit of Liability/Disclaimer of Warranty: While the publisher and author have used their best efforts in preparing this book, they make no representations or warranties with the respect to the accuracy or completeness of the contents of this book and specifically disclaim any implied warranties of merchantability or fitness for a particular purpose. No warranty may be created or extended by sales representatives or written sales materials. The advice and strategies contained herein may not be suitable for your situation. You should consult with a professional where appropriate. Neither the publisher nor the author shall be liable for damages arising herefrom.

For general information about our other products and services, please contact our Customer Care Department within the United States at (800) 762-2974, outside the United States at (317) 572-3993 or fax (317) 572-4002.

Wiley publishes in a variety of print and electronic formats and by print-on-demand. Some material included with standard print versions of this book may not be included in e-books or in print-on-demand. If this book refers to media such as a CD or DVD that is not included in the version you purchased, you may download this material at http://booksupport.wiley.com. For more information about Wiley products, visit www.wiley.com.

Library of Congress Cataloging-in-Publication Data:

Syed, Asif.
    Advanced building technologies for sustainability / Asif Syed.
      p. cm.
    Includes index.
    ISBN 978-0-470-54603-1 (cloth); 978-1-118-24121-9 (ebk); 978-1-118-24127-1 (ebk); 978-1-118-25973-3 (ebk); 978-1-118-25980-1 (ebk); 978-1-118-26019-7 (ebk)
    1. Sustainable buildings. 2. Sustainable design. 3. Building—Technological innovations. I. Title.
    TH880.S94 2012
    720'.47—dc23
                                                                           2011036328

Printed in the United States of America

10 9 8 7 6 5 4 3 2 1

This book is dedicated to my father, S. A. Sattar.

ACKNOWLEDGMENTS

Thank you to my family, Miraj, Azhad, and Rabia for your understanding, support, and patience.

Thank you to my mentors and supporters, M. M. Mohiuddin, Erv Bales, Thomas Gilligo, Marc Lorusso, Peter Flack, Norman Kurtz, Alan Zlotkowski, Lenny Koven, and Paul Bello.

Thank you to my editors, Kathryn M. Bourgoine, Doug Salvemini, and Danielle Giordano.

CONTENTS

INTRODUCTION   viii

Chapter 1   SUSTAINABILITY AND ENERGY   1
   Quality of Life Benefits   7
   Finite Fossil Fuel Resources   8
   Greenhouse Gases   10
   Profits and Savings from Energy Efficiency   11
   Site-to-Source Effect   12
   New LEED Version 2009   13
   Per Capita Energy Consumption   14
   Building Energy End-Use Splits, People Use Energy   15
   Carbon Footprint   17
   Funding Opportunities   19

Chapter 2   RADIANT COOLING   21
   History   21
   Introduction   23
   Why Radiant Cooling?   26
   Applications   28
   Radiant Cooling and Historic Preservation   39

Chapter 3   DISPLACEMENT VENTILATION   41
   History   41
   Introduction   42
   Conventional or Mixed-Air Systems   42
   Difference Between Displacement and Underfloor Air Distribution (UFAD)   47

Applications   48
Large Public Spaces (Cafeterias, Dining Halls, Exhibit Spaces)   48

Chapter 4   CHILLED BEAMS   61
Principle of Operation and Technology   62
Benefits of Chilled Beams   63
Types of Chilled Beams   67
Chilled Beam Applications   72
Chilled Beam Use with Underfloor Air Distribution (UFAD) Applications   78

Chapter 5   UNDERFLOOR AIR DISTRIBUTION (UFAD)   83
Validation of UFAD Designs with CFD Analysis   91
Impact on Buildings   95

Chapter 6   DISPLACEMENT INDUCTION UNITS (DIU)   101
Benefits of Displacement Induction Units   103
History of Induction Units   106
Applications   107

Chapter 7   HIGH-PERFORMANCE ENVELOPE   115
Engaging and Nonengaging Envelopes   116
High-Performance Envelope Definition   117
Most Common Energy Codes: ANSI and ASHRAE 90.1   118
Glazing Characteristics   123
How to Exceed the Mandatory Code Performance   128

Chapter 8   THERMAL ENERGY STORAGE   145
Renewable Energy Storage   146
Conventional Air Conditioning Systems   153
Nonrenewable Energy Storage   156

Chapter 9   SOLAR ENERGY AND NET-ZERO BUILDINGS   163
Net-Zero Step 1: Harvesting Solar Energy   166
Solar Energy in Net-Zero Buildings   177

Net-Zero Step 2: Improve Energy Efficiency of the Building and Its Mechanical and Electrical Systems   181

Net-Zero Step 3: Reduce Consumption   183

Chapter 10 GEOTHERMAL SYSTEMS   185

Introduction   185

Geothermal Heat Pumps   190

Types of Heat Pumps   198

Chapter 11 COGENERATION   205

Other Applications of Cogeneration   207

Cogeneration Technologies   211

Micro-Cogeneration or Combined Heat and Power (Micro-CHP)   221

Chapter 12 DATA CENTER SUSTAINABILITY   223

History of Data Centers   224

2011: Top Ten Trends in Data Centers   225

Power Usage Effectiveness (PUE)   226

Technologies That Can Benefit Data Center Efficiency   230

Office Building Applications   234

Air Management in the Data Center   237

INDEX   239

INTRODUCTION

IT HAS BEEN A HUMBLING EXPERIENCE for me to be part of several high-profile projects in the United States and internationally. Most of these projects had some form of a different approach than conventional systems and almost all of them involved integration between different disciplines of the building design. After completing the projects, some of which were very high profile and received a lot of media publicity, I was approached by building industry professional organizations to speak about the projects. When I did so, it came as a big surprise to me that most people in the industry were not familiar with the new and advanced technologies available. Most people who attended these simple lectures were very curious. The most common question was how they could implement these technologies in their projects. Though most of the technologies were basic, they were different from the conventional industry standards. I saw a great desire in all sectors of the building industry to learn these new and advanced approaches and technologies and implement them in their projects. The problem I saw was that different sectors of the building industry required different levels of information or details about these technologies. It was important for architects to integrate these new and advanced technologies into buildings. The contractors were interested in the availability of materials and products, and in how much they cost, compared to the conventional approach. The owners, building developers, and users wanted to make sure the technologies worked and that the associated costs were justified. A common question: Was the pay back sufficient to offset the savings in energy? The engineers were concerned about the liabilities of trying out new systems and were curious about how to perform the calculations, which they had not been taught, and which were not available in most books or software. To a great degree, I saw that most of the building professionals acknowledged the benefits.

The challenge and opportunity I faced was to write a book that would be beneficial to all sectors of the building industry. The information it contained must not overwhelm any one sector or be too little for another who wants to implement these technologies in their projects. I have tried my best to reach an optimum balance of information, neither too much nor too little. Drawing on my thirty years of experience of working with contractors, construction managers, project managers, owners, architects, end users, and equipment vendors, I have tried to do my best to balance out the information. The other challenge I had was to get this information out as soon as possible. With the ever-increasing demand for sustainability and energy efficiency, time was not there. Almost all projects have some form of sustainability element such as LEED certification, energy use reduction, high-efficiency products, high-performance buildings, and so forth. AIA's adoption of the 2030 goal to make building carbon neutral by 2030 demonstrates the urgency for the need of this information.

Here's how it all started: There was—and now is, more than ever—a need in the building industry to reduce energy consumption. This need is driven by several factors, such as sustainability, reduction in operating costs, and the desire to obtain LEED certification. This situation drives the building industry to new challenges to beat the benchmark. The conventional systems in the building industry, especially in the HVAC sector, are so common that they have become the benchmark for measurement of energy, as established by several codes and standards. Any technology that exceeds the benchmark in reducing energy can be considered as advanced and improved. This standard, which establishes that the minimum performance for energy is ASHRAE 90.1, is used by most states in the United States, and by the United States Green Building Council for LEED certification of buildings, and internationally by several countries. Sustainable buildings aspire to reduce energy by 15 to 60 percent more than required by this benchmark. Fifteen percent better than Standard 90.1 is the minimalist approach, prescribed by the USGBC for LEED certification, and can be achieved with relative ease. Sixty percent better than Standard 90.1 is challenging, but can be achieved by incorporating some of the technologies in this book.

The usual systems are so entrenched in the conventional way of doing business that any change is difficult. The entire building industry—the users, architects, engineers, owners, contractors, product manufacturers, and so forth—is so used to working with conventional systems that a change throws them off and can cause difficulties for all. In order to eliminate these difficulties, a new approach—integrated design—is recommended. This new approach of integrated design is holistic and interconnected; it combines the synergies of all parties. It brings together all parties and stakeholders involved from the beginning of the project. This facilitates buy-in by all parties and eliminates surprises. The bigger challenge to new technologies or solutions is less in technical aspects than in the process, because of the technologies' unknown nature. The unknowns, when discussed up front, will educate and inform all parties involved in the project and make the process easier for all concerned. This can significantly reduce the cost of building. Generally, there is an increase in the cost of anything that is new and not familiar, even if it is much simpler and/or uses fewer materials to build than the conventional method. An integrated approach is strongly recommended for projects where new technologies are to be implemented.

Over the last few years, there has been significant publicity, in the media of the building industry and in the general media, highlighting the need for sustainable and energy-efficient buildings—but a lot still has to be done. Buildings are essential to the life and work of all human beings and benefit us in enormous ways. The environmental impact of buildings has become a significant factor in their design and construction. As per data published by United States Environmental Protection Agency, buildings in the United States use about 40 percent of the total energy produced, consume 68 percent of the total electricity produced, and account for about 38 percent of carbon emissions. The environmental impact of buildings is significant, and any steps taken to reduce their energy and electricity consumption have the dual advantage of both environmental and economic benefits. Any energy or electricity not consumed is money not spent on fuel. Reduction in the operating costs from consuming less energy can only benefit the competitiveness of the business in the global market. This book gives examples of projects that have implemented the new technologies to reduce energy consumption and contribute to the sustainability goals of the buildings.

The sustainable technologies can be evaluated based on their own merits of return on investment and life-cycle costs. Most building systems have a long operating life. In older buildings, systems installed as far back as forty years are still operating. The new energy saving technologies will also have a similar life cycle. The long building life cycle leads to the high rate of return on investment. The payback on advanced technologies can vary from as low as two to as long as twenty years. A yearly cash flow analysis for the life cycle is the best way to demonstrate the rate of return on the investment. Most sustainable technologies support a rate of return due to their long-term use over fifteen to twenty years. Readers are encouraged to investigate these and understand the financial issues prior to working with the technologies. The financial return on investment analysis is the best tool to convince the critics of sustainability.

The technologies are constantly evolving with innovations in construction products and materials, lessons learned from operating, new system designs, and construction means and methods. The technologies in the book are not the end of the line toward sustainability, but only a beginning. Our experience and information gathered will make these systems better and more efficient. These technologies are the first ones to replace the conventional systems. As they get more widespread the costs will reduce and improvements will be made, making them more efficient.

CHAPTER 1

Sustainability and Energy

BUILDING ENERGY CONSUMPTION IS A SIGNIFICANT PORTION of the total energy used worldwide. In the United States, buildings use about 40 percent of the total energy consumed and about 68 percent of the electricity produced. Buildings are responsible for 38 percent of carbon emissions.[1] Buildings account for the highest carbon emissions, followed by transportation and industry. Buildings will continue to grow as the population of the world grows. The current world population according to the U.S. Census Bureau is 6.9 billion[2], and is projected to grow from 6.1 billion in 2000 to 8.9 billion in 2015.[3] The growth in population creates demand for new buildings: residential, educational, commercial (office and retail), health-care, and manufacturing. Growth of the buildings is going to happen, whether we like it or not. These new buildings will increase the demand for energy, increasing the cost of energy. Additionally, the growth of buildings will increase the global carbon emissions.

Economic development is essential to the social, political, and economic order of the world, and building construction is a big part of the economic development of all the world's countries. With almost 9 million people employed in construction (per 2008 statistics), it is one of the largest industries. The wages of construction workers are relatively high. The construction industry also creates and promotes small business, as more than 68 percent of construction-related establishments consist of fewer than five people, and a large number of workers in construction are self-employed.[4]

---

[1] National Institute of Building Sciences, Whole Building Design Guide, 2009.
[2] US and World Population Clock, US Census Bureau, www.census.gov/main/www/popclock.htm.
[3] World Population to 2300, Department of Economic and Social Affairs, United Nations, 2004.
[4] United States Department of Labor, Bureau of Labor Statistics, www.bls.gov/home.htm.

Figure 1-1 Building energy use according to the U.S. Department of Energy. *Buildings Energy Data Book*

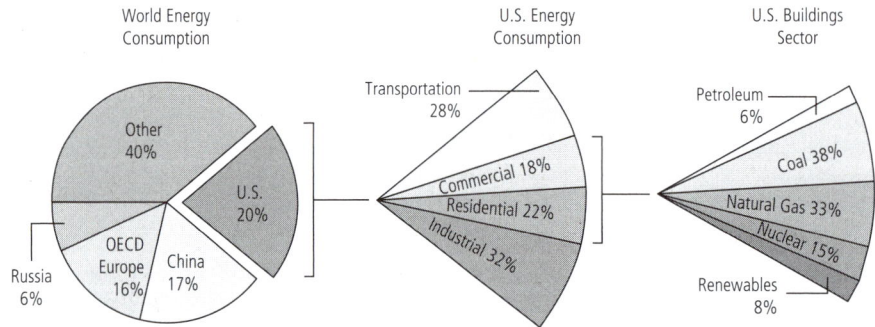

Economic development and growth will continue to add new buildings. The new buildings present an opportunity to adopt new technologies and reduce the increase in demand for energy, thus containing the cost of energy. Slowing down the increase in energy consumption through advanced technologies also reduces carbon emissions, reducing the impact of development on the environment.

In the 1300s, Arab historian Ibn Khaldun defined or described economic growth as:

When civilization or population increases, the available labor or manpower increases. In turn, luxury increases in correspondence with the increasing profit, and the customs and needs of luxury increase. Crafts are created to obtain luxury products. The value realized from them increases, and, as a result, profits are again multiplied. And so it goes with the second and third increase. All the additional labor serves luxury and wealth, in contrast to the original labor that served the necessity of life.

Versions or parts of Ibn Khaldun's theory are still valid in modern times, which means that economic development is imminent and ongoing. Construction of new buildings is a big part of economic development and will continue, as a result of:

- Growth due to increase in population
- Higher rate of growth in the developing countries due to globalization
- A very high disparity between the per capita energy consumption and building footprint in developing countries vs. developed countries
- Trying to catch up with developing countries puts additional demand above and beyond normal population growth.

New technologies can contribute to slowing down the growth of energy consumption, without slowing down the economic growth that is essential to maintain the world's social, political, and economic order. The goal of energy savings in buildings is to reduce the rate of growth of energy consumption, while maintaining economic growth. World economic growth is expected to grow 49 percent by 2035, as reported by the United States Energy Information Administration report *International Energy Outlook 2010*.[5]

Continuing at this rate of growth and development with the current practices of using energy, which primarily comes from using fossil fuels, has two diametrically opposite forces. On one side is growing more, traveling more, having more space, and brighter and bigger cities. On the other side, there are limited or declining resources. Effectively utilizing resources is essential or soon it will take more than one earth to meet the growing needs for resources. "Soon" is now, according to the

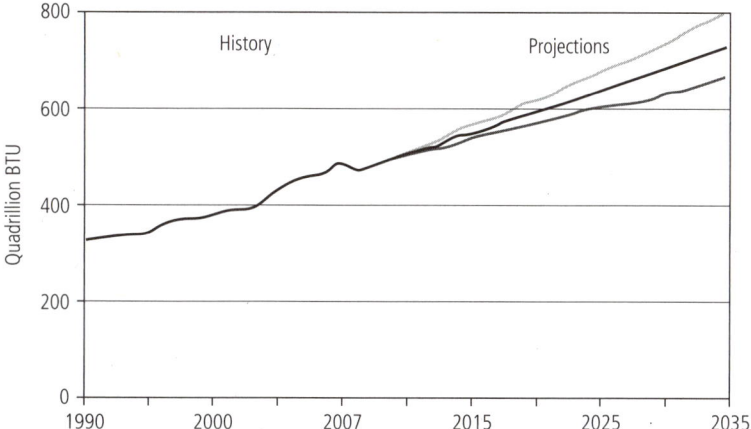

Figure 1-2 World marketed energy consumption in three economic growth cases, 1990–2035. *U.S. Energy Information Administration*

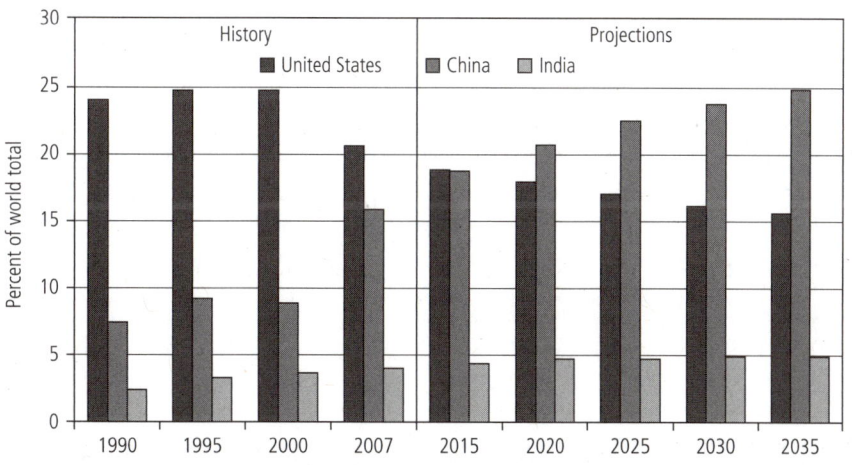

Figure 1-3 Shares of world energy consumption in the United States, China, and India, 1990–2035. *U.S. Energy Information Administration*

---

[5] United States Energy Information Administration report, *International Energy Outlook 2010*, http://38.96.246.204/forecasts/ieo/".

Global Footprint Network, an alliance of scientists who calculate that in 10 months, humanity will have exhausted nature's yearly budget.[6]

Growth in the developing countries will occur at a much higher rate than in the developed Western world. The U.S. Energy Information Administration has made two forecasts: high economic growth (63%) and low economic growth (37%). With higher growth rates in the developing world or the newly industrialized countries, the median growth of 50 percent is very likely, based on the growth rate and energy consumption growth of India and China. To a certain degree, India's and China's energy consumption growth will not put all the pressure on fossil fuels, given their high level of growth in nuclear power plants. According to the World Nuclear Association,[7] nuclear power generation has the highest growth in Asia. China, Japan, South Korea, and India are the countries with the largest number of nuclear power plants planned; more than eighty-four nuclear power plants are planned in these countries. The recent tsunami in Japan has exposed the vulnerability of nuclear power generation. The damages from the tsunami are evident, and several countries are reevaluating their dependence on nuclear power. Every country is evaluating whether the benefits are worth the risks. It is too soon to predict (1) the pressures that the increase in demand for energy will put on the prices of clean-burning fossil fuels or (2) the huge environmental impact of growth in conventional coal power plants.

The worldwide economic growth will put intense pressure on energy resources and will increase the demand for energy and fossil fuels. In the current methodology of energy production, energy and fossil fuels are almost synonymous, as currently fossil fuel is the major source of energy. Fossil fuels such as oil, coal, and natural gas account for more than 85 percent of the energy used in the United States.[8] The same fossil fuels produce about 70 percent of the electricity. The other 30 percent breaks down as 20 percent from nuclear power plants, 6 percent from hydro power plants, and 4 percent from renewables such as solar and wind power.[9] Making improvements to buildings' energy use and efficiency can generate significant savings in energy and fossil fuel costs. The majority of fossil fuels are a globally fluid commodity that flows to the place of demand or to the highest bidder. The fluidity of the fuel creates a global demand. The increase in demand is much higher in the developing countries.

Potential opportunities exist to make improvements to buildings in the mechanical, electrical, and plumbing (MEP) systems to improve their energy efficiency. There are currently available technologies that are cost effective and can reduce energy

---

[6] "Global Footprint network," www.footprintnetwork.org/en/index.php/GFN/page/earth_overshoot_day/.
[7] "Asia's Nuclear Energy Growth," World Nuclear Association, April 2010.
[8] United States Department of Energy, www.fe.doe.gov/index.html.
[9] United States Energy Information Administration, Electric Power Monthly report, released Nov 16, 2011. Table 1.1, "Net Energy Generation by Source," 2011, www.eia.gov/electricity/monthly/index.cfm.

consumption by a significant amount. According to guidelines published by the American National Standards Institute (ANSI), the American Society of Heating, Refrigerating and Air-Conditioning Engineers (ASHRAE), and the Illuminating Engineering Society of North America (IESNA), almost 50 percent of energy can be reduced in office buildings.[10] The most common standard used the world over, and adopted by most states in the United States is the ANSI/ASHRAE/IESNA Standard 90.1. The same professional organizations that wrote Standard 90.1 have also written design guides on how to achieve up to 50 percent energy savings over their own standards. Clearly, from these publications, there is evidence that there is significant opportunity to reduce energy in buildings. According to the United States Green Building Council (USGBC), a nonprofit organization that promotes sustainability in the building industry, there are potential technologies for existing and new buildings that can reduce energy use by 25 percent and carbon emissions by 30 percent.[11] Moreover, there are opportunities to continue with growth and its economic benefits, but reduce the impact on energy resources, fossil fuels, and the environment by adopting the efficient technologies.

However, these technologies are not commonly known to the construction industry, including most design professionals, contractors, and manufacturers of building construction equipment and materials. Most of these new and advanced technologies or design approaches are basic and simple in nature, and easily understandable and implementable. However, they are different from the current popular practices employed by the building design industry, including design professionals, contractors, and building operators. There are a select few professionals, both architects and engineers, who are familiar with and can confidently design these new technologies or mechanical or electrical systems; however, for the majority of the construction industry, these are technologies they have only heard about or read about. The "unknown technology factor" is the biggest barrier to the use of the more efficient and advanced technologies. To a certain degree, the problem is also the need to break an old habit or to change "business as usual." To successfully implement the new and advanced solutions, there has to be a change in attitude, approach, and practice in the profession. This change is very difficult to bring about in a well-established industry such as the building construction industry, which is a major contributor to the overall economic activity of the United States and the rest of the world. Construction totals to about $800 billion a month, resulting in approximately $9 trillion per year.[12] The construction industry is one of the largest and is well set in its systems, practices,

---

[10] *Advanced Energy Design Guide for Small to Medium Office Buildings: Achieving 50% Energy Savings Toward a NetZero Energy Building*, ASHRAE.
[11] United States Green Building Council, www.usgbc.org/.
[12] "Value of Construction Put in Place—Seasonally Adjusted Annual Rate," United States Census Bureau, www.census.gov/const/www/c30index.html.

methods, and approach. Even a small change in this industry is difficult and takes a long time. However, there are positive trends; many projects that have incorporated advanced sustainable technologies are featured in the press and have received positive publicity with their success. Professional organizations such as the American Institute of Architects (AIA) and ASHRAE are promoting these technologies. Government bodies such as the Department of Energy (DOE) are promoting energy efficiency with several programs such as Portfolio Manager, whereby buildings are ranked by their energy consumption compared to similar buildings. Only five to six years back, the universal answer of builders and designers to the question, "Does it cost more to adopt sustainable technologies?" was, "Yes." Now, many builders and designers—if not all—can confidently say, "It does not cost more to employ sustainable technologies." This is a significant shift in position over the last five years. Also, most building owners have voluntarily adopted sustainable technologies to reduce energy use or to be green. Most building owners are designing and operating buildings to USGBC, to obtain LEED certification.

The cost of these new or advanced technologies is not necessarily higher than that of the conventional systems. However, it depends on whom you ask. Professionals who are familiar with the advanced systems will agree that the construction cost is the same, and that if there is an additional cost, it usually is recouped within a reasonable payback period. Professionals who are unfamiliar with these systems will generally believe that advanced technologies cost more, primarily because the "unknown technology factor" raises the cost far higher than the true cost. Some of these technologies do not cost more than conventional systems; they are simply different. Some may cost more for one item, but reduce costs for other items. For example, in underfloor air conditioning systems, the cost of the raised floor is higher, but there is no need to install ductwork and associated accessories such as variable air volume (VAV) boxes and the like. If there are any additional costs, usually they have a very short payback period. The increase in the cost is offset by the energy savings. The payback is calculated with energy analysis and life-cycle cost analysis. Life-cycle cost analysis has not been part of the construction industry; most design professionals are unfamiliar with it. Thus, these professionals are not able to calculate the necessary life-cycle cost or yearly operating cost to demonstrate how payback will justify the expense. Lack of knowledge of or familiarity with new and advanced technologies is limiting. Most in the construction industry tend to stay with what they know and have experience with. This book will demonstrate that the new technologies are basically energy efficient, sound, simple, easy to build, user- and operator-friendly, and cost effective. It will be a small step toward making the entire construction industry familiar with new and different solutions, which will eventually remove the fear of the unknown. This knowledge and awareness in the building construction and operations industry will transform the way buildings are designed, built, and operated.

QUALITY OF LIFE BENEFITS

In addition to their energy and environmental benefits, new technologies improve the quality of life for a building's occupants. Indoor air quality is one of the major factors that affect the quality of life in buildings. People spend 90 percent of their time inside buildings, making it all the more important to focus on the quality of life a building provides for its occupants. Sick building syndrome (SBS) explains why those who spend a lot of time in a building complain of ill health and discomfort, with no apparent cause. The causes of sick building syndrome are generally:

1. The growth of bacteria and molds in the buildings, due to inadequate temperature and humidity control
2. Inadequate ventilation, which is affected by the amount of outside air introduced into the building
3. Ineffective ventilation, which generally results when outside air introduced into the building bypasses the occupants
4. Indoor chemical pollution from off-gassing of building materials and finishes, such as volatile organic compounds (VOC)

Advanced systems, in addition to reducing energy use, have better indoor air quality than conventional systems, leading to better health for the occupants. Indoor air quality is just as important as outdoor air pollution—and in some instances more important. Since people spend 90 percent of their time in buildings, indoor air quality is an important factor in their well-being. Most of the conventional systems that are predominant in the building industry do not improve indoor air quality, and in most instances are detrimental to it. The EPA recommends three basic strategies for improving indoor air quality:[13]

1. Source control
2. Improved ventilation
3. Air cleaners

Two out of the three recommendations are systems-related. Improved ventilation can be achieved by increasing the percent of outside air that is circulating in the building. Only from 15 to 20 percent of the total air circulating in a typical building is outdoor air; 85 percent is recirculating air. LEED certification recognizes this, and in their point-based rating system, the USGBC provides means of increasing ventilation and achieving additional points. However, while implementing this process, careful analysis

---

[13] "An Introduction to Indoor Air Quality," EPA. www.epa.gov/iaq/ia-intro.html.

has to be made to evaluate the outdoor air quality level. Some areas in the country, especially urban environments, have high levels of outdoor contaminants. Another common way to improve ventilation is by providing operable windows, enabling building occupants to decide the need for more outside air. Improved ventilation may not necessarily result from an increase in outside air, but from the effectiveness with which the outdoor air is delivered to occupants. The conventional systems really fall short in delivering outside air to occupants effectively. The standard overhead air distribution systems mix pollutants in the air delivered to a space, increasing the parts per million (PPM) of contaminant particles at the occupant breathing elevation. Advanced systems such as underfloor air distribution (UFAD) systems reduce the PPM of contaminant particles.

The most effective way to keep the indoor building environment or air clean is to control the source. The source of the contaminants can be indoors or outdoors. Inside source control is accomplished relatively easily by properly selecting the materials and furnishings that make up the indoor environment. Huge strides have been made in this sector, and most indoor materials are rated or labeled with their potential emission of contaminants. Increasing the outdoor air percentage is another way to control indoor source pollution, as the outside air will dilute the contaminants. Increasing the outdoor air percentage of the circulating air has some limitations, however. Depending on the location of the building, the outdoor environment may be too hot or too cold, requiring excessive energy to heat or cool the outside air. Some regions may have harmful levels of outdoor contaminants, limiting the amount of outside air use. Therefore, a good air-cleaning system is essential. Improvement of the efficiency with which the air-cleaning system captures the contaminants from the circulating air is essential in both conventional and advanced systems. LEED building-rating systems recommend a minimum efficiency reporting value of 13, or MERV –13, for permanently installed mechanically ventilated systems, for circulating both building air and outside air. An air-cleaning system can be detrimental to the overall system, however, because a fine filter or air cleaner requires additional energy. Indoor air quality control is a balance of several variables that include: indoor contaminants, outdoor contaminants, ventilation effectiveness, the outdoor environment, filtration, and the delivery system.

FINITE FOSSIL FUEL RESOURCES

Most of the energy we produce and consume comes from finite resources. About 56 percent of the energy produced in the United States comes from finite resources such as coal (22%), natural gas (21%), crude oil (11%), and natural gas liquids (3%). All the resources are finite and will not last forever. Even coal, the largest energy reserve,

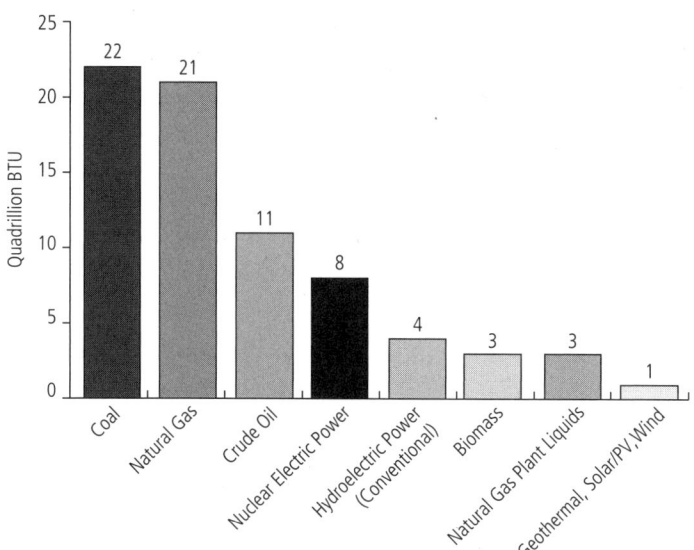

Figure 1-4 U.S. Primary Energy Production by Major Source (2009). *U.S Energy Information Administration, Annual Energy Review, 2009, Table 1.2 (August 2010)*

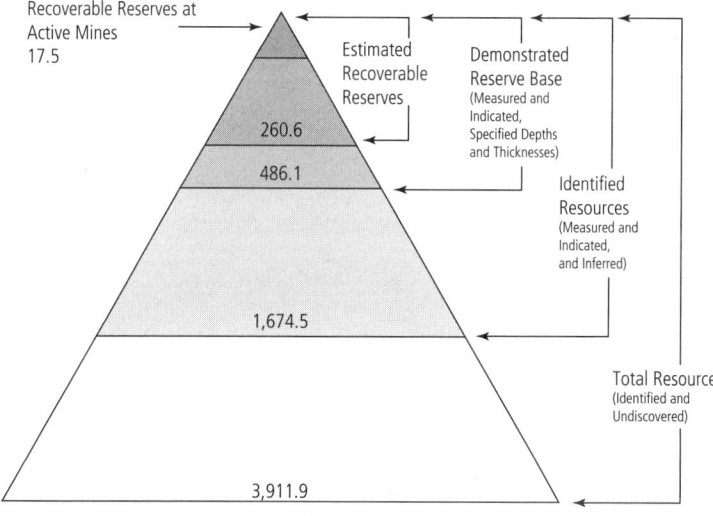

Figure 1-5 U.S. Coal Resources and Reserves (Billion short tons as of January 1, 2010). *U.S. Energy Information Administration, Form EIA-7A, Coal Production Report (February 2011)*

is a finite resource.[14] The very definition of sustainability, "endure without giving way or yielding," conflicts with the use of finite resources such as fossil fuels, which will run out eventually. Until alternate nonyielding resources are tapped into, or technologies are developed to fully utilize renewable resources, it is essential to reduce energy consumption by buildings. A 30 to 50 percent reduction in the energy consumption of buildings can lead to a 12 to 20 percent reduction in overall energy use. The ultimate

---

[14] U.S. Energy Information Administration, U.S. Coal Resources and Reserves, 2010.

goal is to have all energy come from renewable sources such as wind, geothermal, and solar power. But this will not come about in the immediate or near future. The current focus is on reducing energy consumption by improving the efficiency of building systems, which will accelerate the ultimate goal of relying exclusively on renewable energy.

GREENHOUSE GASES

Gases that trap heat from the sun are called greenhouse gases. These gases are essential to life on the Earth in its current form. It is the greenhouse gases that maintain the temperature on the Earth that sustains life. In the absence of the greenhouse gas effect, the temperature of the Earth would be lower by 60°F.

There are several greenhouse gases—the six identified by the U.S. Energy Information Administration and the Kyoto Protocol are:

1. Carbon dioxide (CO_2)
2. Methane (CH_4)
3. Nitrous oxide (N_2O)
4. Hydrofluorocarbons (HFCs)
5. Perfluorocarbons (PFCs)
6. Sulfur hexafluoride (SF_6)

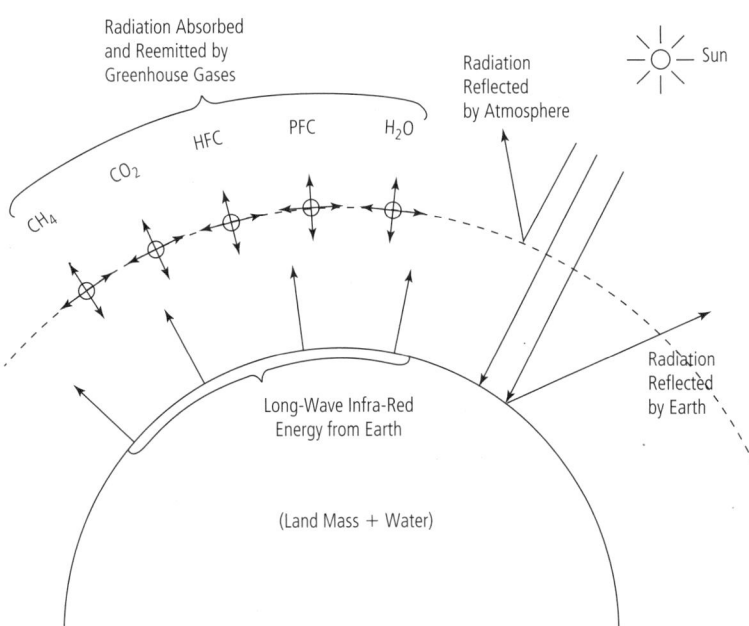

Figure 1-6 Greenhouse gas effect. *Asif Syed*

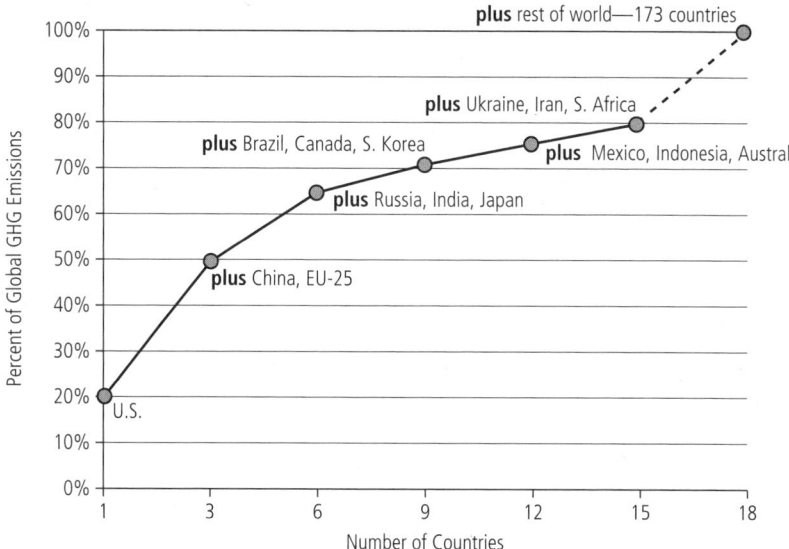

Figure 1-7 Aggregate Contributions of Major GHG Emitting Countries. *U.S. Energy Information Administration*

When sunlight strikes the Earth, some of the energy is re-radiated back into space as infrared energy. All greenhouse gases absorb this re-radiated energy as infrared radiation (heat). The absorbed energy of the greenhouse gases causes heat to be trapped in the atmosphere. Burning fossil fuels leads to the production of carbon dioxide (also referred to as CO_2) emissions. Of the listed greenhouse gases, carbon dioxide is the largest contributor to the greenhouse gas effect. Advanced systems reduce the consumption or burning of fossil fuel for energy, thus reducing the production of carbon dioxide. This leads to the reduction of greenhouse gases.

PROFITS AND SAVINGS FROM ENERGY EFFICIENCY

Energy savings have a direct impact on the bottom line of businesses and building owners. Energy saved through conservation measures and efficiency is energy not consumed. The unconsumed or saved energy does not have to be paid for. The savings from energy efficiency are not commonly discussed. Because of the low cost of energy in the past, compared to the overall or total cost of operating a building and business, the energy budget was small compared to the overall budget of business operation. The cost of energy was so small that it did not stand out or constitute an important factor. This is similar to the cost of gasoline for cars—it was not common to calculate the cost of gasoline while using a car. However, the state of the economy after 2007—with a recession second only to the Great Depression of the 1930s—has caused scrutiny on these aspects of business costs. The higher cost of energy prior to the recession and the apparent waste of energy have put the focus on energy savings and the costs associated with energy use. Savings from energy conservation and efficiency are directly related and proportional to energy saved and greenhouse gases reduced. This has been demonstrated at St. John's University, in New York City, which saved $1,100,000 in operating costs while reducing greenhouse gases by 9,270 metric tons' equivalents of carbon dioxide.[15] St. John's University has an energy efficiency improvement program and participates in the carbon footprint reduction program called the 3010 challenge. The 3010 challenge is for the educational institutions in New York City who voluntarily participate in the program, to reduce their carbon footprint by 30 percent in 10 years. The 3010 challenge is part of New York City Mayor Michael Bloomberg's program to reduce the carbon footprint of New York City, called "Plan NYC."

SITE-TO-SOURCE EFFECT

The amount of energy used in buildings as measured by electric utility meters, natural gas meters, or the measure of fuel oil delivered is not a true representation of the energy consumed by the building. The amount of energy generated at the power plant is much higher. This is especially significant for electricity and is about three times that used at site or in the building. For a 100 watt LED TV, about 300 watts of equivalent fossil fuels has to be burned in the power plant. Site energy is the amount of energy consumption reflected in the utility bills, but it is not the true representation of energy use. The primary form of energy bought at the building site, such as natural gas, comes from a distant location, and losses are associated with it. The most common form of energy used in buildings is electricity, which is considered a secondary form. Electricity is produced by burning a fossil fuel or by a hydro or nuclear power plant, but the most common form of fuel for electricity is fossil fuel. The secondary form of energy electricity is produced in a power plant. Most thermal power plants have only about 30 percent efficiency. So the energy equivalency is much larger at source than at site. In the case of electricity—the most common form of energy—the site energy equivalency is about 3.34.[16] The site-to-source factor includes the thermal efficiency losses and transmission losses. For the most common energy uses, the EPA methodology for calculating site-to-source conversion factors is as follows:

---

[15] St. John's University, Environmental Assessment Statement: Memorandum of Understanding, Semi-annual Report, July 2011.
[16] Energy Star performance ratings, methodology for incorporating source energy use.

TABLE 1-1 SITE-TO-SOURCE CONVERSION TABLE

#	Fuel Type	Site-to-Source Ratio
1	Electricity (grid purchased)	3.34
2	Electricity produced on-site from solar or wind	1.0
3	Natural gas	1.047
4	Fuel oil	1.01

Generation of electricity with fossil fuels is a very inefficient process, with losses as high as 60 to 70 percent. The losses are in the form of heat in the flue gases of the combustion, which are vented into the atmosphere. The heat from the flue gases is not useful in most locations of the power plants, which leads to lower efficiency. Site-to-source conversion is especially important because any reduction in energy at site is almost three times the energy saved at the power plant. The reduction in the greenhouse gases is also three times the amount of energy saved.

NEW LEED VERSION 2009

The new LEED rating system has increased the emphasis on energy. When United States Green Building Council's LEED rating system started, certification—or higher levels such as silver or gold—did not require mandatory points in the energy and atmosphere category. However, in 2007, for basic certification or higher ratings, the emphasis on energy increased, and it became mandatory to obtain two points in energy and atmosphere. Two points meant 14 percent better than the energy code minimum. This has forced architects and engineers to come up with innovative designs. The standard used for energy code minimum is almost universally the American Society of Heating Refrigeration and Air Conditioning Engineers ASHRAE 90.1. The ASHRAE 90.1 standard is becoming more and more efficient as newer versions are introduced every three years. Achieving lower than baseline minimum code was much easier in the past, but with newer versions it is more challenging. Some or most of the advanced technologies are still not the minimum or baseline code requirements, presenting an opportunity to exceed mandated energy efficiency and add more points toward LEED certification.

USGBC's LEED certification process is continuously increasing its emphasis on energy. In the earlier versions of certification, the only energy prerequisite was to comply with code. Additional energy use reduction was optional. Later, in 2007, a mandatory rating of 14.5 percent better than the code became a prerequisite for

TABLE 1-2 LEED RATING SYSTEMS ENERGY OPTIMIZATION POINTS

LEED Rating Version for New Construction	LEED Energy/Total Points	% of Total Points
LEED 2	10/69	14.5%
LEED 2.1	10/69	14.5%
LEED 2.2 (before July 2007)	10/69	14.5%
LEED 2.2 (after July 2007)	10/69 – 2 mandatory	14.5% (2.8% mandatory)
LEED 2009	19/110	18.2%

certification. The LEED 2009 rating system has increased the importance of energy by increasing the points for energy credits. LEED 2009 has 19 points in a 100-point system, with an almost 20 percent emphasis on energy. To achieve higher ratings such as gold and platinum, advanced systems can be used to maximize the points.

PER CAPITA ENERGY CONSUMPTION

The per capita energy consumption of all countries indicates that there is a big gap between the developed countries and the developing countries. The average power consumption[17] of developed countries is 200 MBtu, whereas in the developing countries it is about 20 MBtu. The developing countries have populations that are much larger than those of the developed countries. With the total population of the world at 7 billion, 6 billion people live in the developing world, and only 1 billion in the developed countries.[18] The huge populations of developing countries aspire to the quality of life and the lifestyles of the developed countries. If the populations of developing countries start consuming the same 200 MBtu, the consumption of energy will not be sustainable. Sustainable technologies can help in lowering energy consumption in the developing world. However, most developing countries are using systems that were used in the developed world in the 1970s and 1980s. The technologies of 1970 and 1980 were not energy efficient. Generally, the developing countries emulate what is being done in the developed world. This cycle has to be broken, and new and advanced technologies have to be adopted in the developing world, alongside the developed world, to make a difference in the overall energy consumption of the world.

---

[17] U.S. Energy Information Administration—International Energy Annual, 2006.
[18] "United Nations Environmental Program—Trends in population, developed and developing countries," http://maps.grida.no/go/graphic/trends-in-population-developed-and-developing-countries-1750-2050-estimates-and-projections.

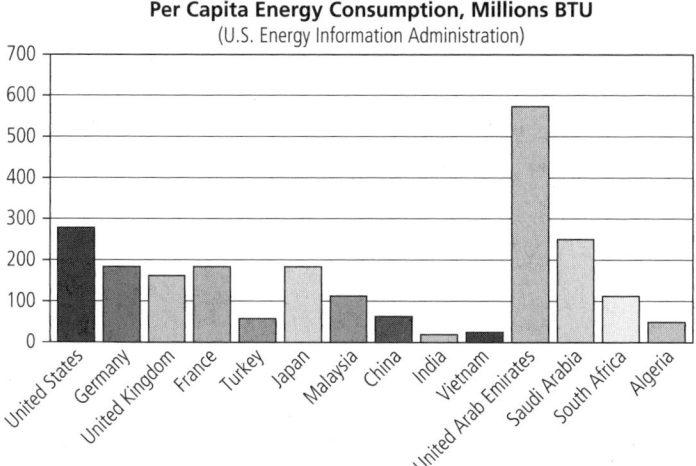

Figure 1-8 Per capita energy consumption of representative countries of the world. *U.S. Energy Information Administration*

BUILDING ENERGY END-USE SPLITS, PEOPLE USE ENERGY

The data collected[19] by the U.S. Department of Energy indicates that almost 50 percent of the energy consumed by buildings goes into serving the occupants' needs, such as water heating (9.6%), electronics (7.6%), refrigeration (5.5%), cooking (3.4%), computers (2.3%), and so on, and that the remaining 50 percent goes into space lighting, heating, and cooling. People use energy whether they are in large commercial buildings, at home, or elsewhere. The space heating, which is 20 percent, provides protection from the elements and is a necessity, whether people are in structured commercial buildings or just at home. Space cooling provides comfort; it was originally considered a luxury, but it has become a universal necessity, and most homes are now air-conditioned.

People consume energy, and building systems are a means of delivering the energy. Reducing energy in buildings is a twofold issue: occupants and systems. The ratio of influence of the people and the system is an equal fifty-fifty split. Actions or behaviors of the building users and occupants can make a significant difference in the overall energy consumption. An integrated design approach of advanced systems engages the occupants and brings about the awareness of the entire building design process and the importance of energy and sustainability. Occupants learn how their behavior has an impact on the building's energy and sustainability. This will lead to change in behavior, which will be a benefit.

---

[19] U.S. Department of Energy—2006 U.S. Buildings Energy End-Use Splits, http://buildingsdatabook.eren.doe.gov/ChartView.aspx?chartID=1.

In the present building design process, the art and science of occupant behavior impact on energy does not exist. Occupant behavior is not considered as a design issue. Integrated design and engaging advanced technologies is a starting point, but soon a new chapter has to be written on this subject. In net-zero buildings, especially the ones when energy is produced on-site with solar photo voltaic cells, occupants understand this. The behavior of occupants can reduce energy, which does not have to be produced in a photo voltaic panel, leading to lowering the capital cost. The initial capital investment associated with solar photo voltaic panels can have an influence on the behavior aspects of the occupants. Some examples of behavior can be using natural light and ventilation.

The efficiency level associated with the systems that deliver the energy to the occupants can be improved with the assistance of the advanced technologies now available. For example: Fans commonly used in buildings are only 65 percent efficient, and air used to transport cooling has extremely low heat-carrying capacity, or specific heat. On the other hand, pumps are 85 percent efficient, and water has very high heat-carrying capacity. Selection of a water-based system can significantly lower the energy consumed. Harvesting daylight by appropriately selecting glazing and lighting control systems, such as dimming, can reduce the lighting energy consumption, which is a significant split. Technologies for glazing include spectrally selective coatings that reduce solar heat gain and maximize light transmission.

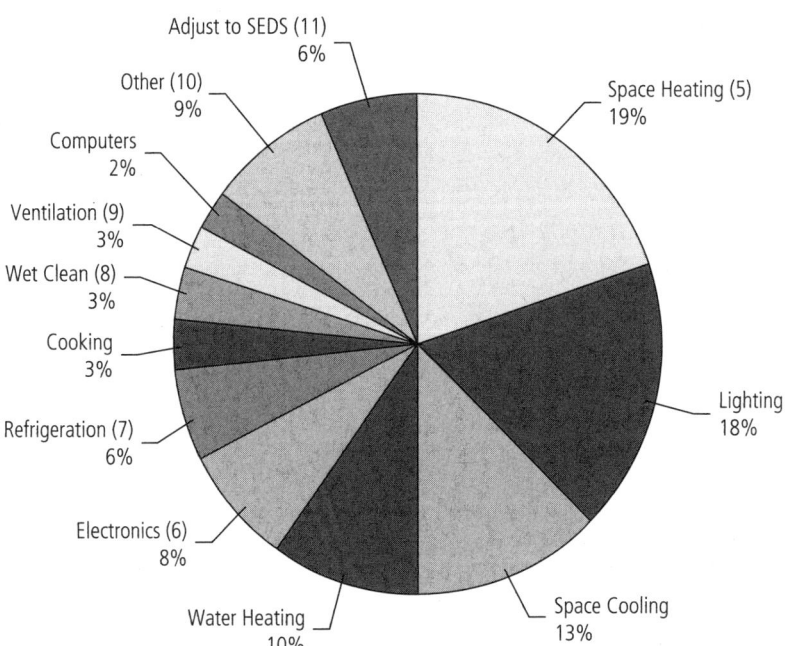

Figure 1-9 U.S. Buildings Energy End-Use Splits. *U.S. Energy Information Administration*

CARBON FOOTPRINT

Most organizations, in both the public and private sectors, are becoming carbon footprint conscious. Carbon Disclosure Rating (CDR) is a numerical score based on the level of reporting of a company's climate change initiatives. This is in response to the questionnaire that was developed by the U.K.- based Climate Disclosure Project (CDP) along with PricewaterhouseCoopers. The score is not indicative of the actions taken by the company to mitigate it's climate change issues. The score only indicates the level of disclosure of a company's climate change issues. A high score generally indicates a good understanding and management of issues that impact the climate from a company's activities. Most large companies have a Carbon Disclosure Rating. Carbon disclosure ratings are given for stocks' symbols along with the companies' profit margins, P/E ratios, and return on assets. Companies that are tracking their carbon footprint and their climate change impacting issues need to equip their building systems with advanced technologies that use less energy and thereby leave a lower footprint reducing their impact on climate change. The carbon footprint is a measure of the release of all the six gases identified by the Kyoto Protocol as greenhouse gases: carbon dioxide (CO_2), methane (CH_4), nitrous oxide (N_2O), hydrofluorocarbons (HFCs), perfluorocarbons (PFCs), and sulfur hexafluoride (SF_6). The carbon footprint of an organization or campus or company is the amount of these six gases released directly or indirectly. The measure of the carbon footprint is in tons of carbon dioxide released into the atmosphere. For the other five gases, the measurement used is the effect of these gases on global warming, calculated as carbon dioxide equivalent. Carbon dioxide is used as the baseline.

EMBODIED ENERGY VERSUS OPERATIONAL ENERGY

For working toward carbon neutral or net-zero buildings, understanding operational and embodied energy of buildings is important. Operational energy is the energy consumed annually by the building MEP systems for heating, cooling, appliances, and lighting. The operational energy is based on the type of MEP systems adopted in the building and is easily measured with meters and estimated prior to design with analytical tools such as computer simulation energy software. Embodied energy is the energy used in mining, manufacturing, transporting, installing, and finally demolishing the materials that are used in the building. Operational energy is the majority of the energy consumed in the building over its life cycle. The embodied energy of different materials vary based on the type of materials used such as concrete or steel or wood. The embodied energy also depends on the transportation of building materials from the harvesting site to factories and to the construction site. Wood harvested from renewable forests provides a sequestering effect. While growing wood carbon is captured from the atmosphere and the energy required to produce wood all comes from the sun, a renewable resource. When the wood is used in long-term application such as a building material with a life of fifty years, the carbon is sequestered for that period. The amount of embodied energy and operating energy varies based on the type of building such as retail, residential, commercial, and so forth. Buildings that operate 24/7 like hospitals use far more energy than office buildings that operate only ten hours per day. The embodied energy and operating energy ratio also depends on the life cycle of the building. As the life of the building increases, embodied energy stays the same, while the operating energy goes up. The operating energy is almost three to four times the embodied energy over the life cycle of the building.

Embodied energy is about 20 to 25 percent[20] of the energy over a fifty-year life cycle of the building, while operational energy is 75 to 80 percent. Operational energy is the energy consumed during the building's life once it has been constructed. This energy is consumed by heating and cooling, lighting, and appliances, which

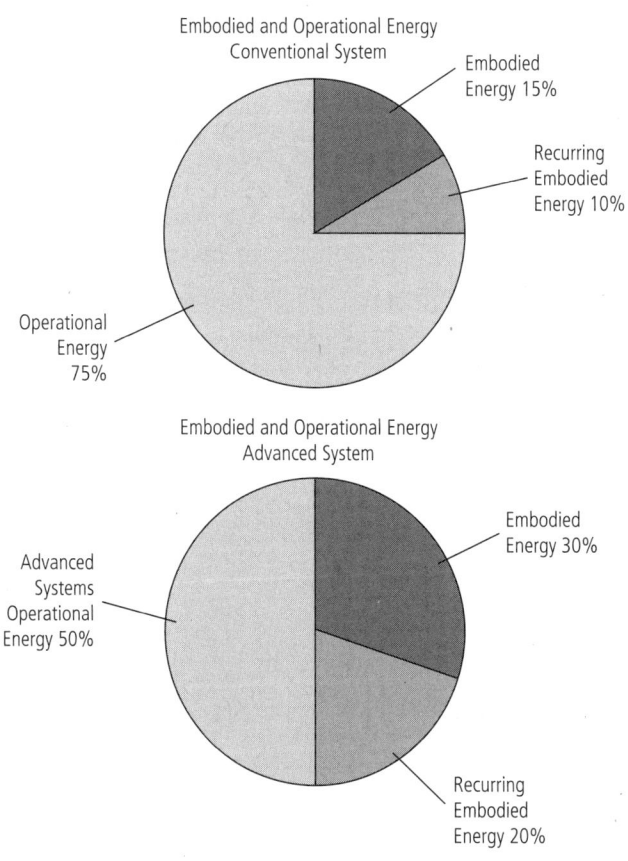

Figure 1-10 Embodied and Operational Energy—Conventional System and Embodied and Operational Energy—Advanced System. *U.S. Energy Information Administration*

---

[20] http://architecture2030.org/about/design_faq#embodiedenergy.

includes the mechanical and electrical systems delivering this energy to the building. The energy efficiency of the building systems can make a significant impact over the life of the building.

FUNDING OPPORTUNITIES

In order to promote the advanced solutions and technologies for sustainable and energy-efficient operations, several grants and funds are offered by federal and state governments, quasi-government agencies, and public utility companies. While several of the technologies have a relatively low payback or return on investment, this is not always the case. For technologies with higher payback, funding opportunities can help in reducing the payback or the return on investment. The American Recovery and Reinvestment Act of 2009 (Recovery Act), and other grants provide funds to the U.S. Department of Energy and other agencies. The Office of Energy Efficiency and Renewable Energy (EERE) have financial assistance programs for the use of renewable energy and energy efficiency technologies. Most of these programs are based on the funding available, and most of the funds are fixed and are sometimes on a first-come first-serve basis. Very early on during the evaluation of these technologies, such opportunities must be investigated. However, it should not be mistaken that for all advanced technologies such funds or grants are required to make them financially feasible. Some of them can work by themselves, while others require assistance. These programs help reduce the long-term production costs of some technologies, especially solar photo voltaic systems, which are primarily driven by substantial assistance from federal, state, and local utility cash rebates or tax incentives. From 1998 to 2010 the average cost of photo voltaic installations has reduced by more than 50 percent. The current costs are in the range of $5 to $6 per watt compared to $12 to $16 in 1998.

A database of the public funding opportunities is available on the Web. The Database of State Incentives for Renewables and Efficiency (DSIRE) was established in 1995 and funded by the Department of Energy. The website lists all state programs currently providing funding, and also lists federal funding programs. DSIRE is run by the North Carolina Solar Center and the Interstate Renewable Energy Council. Funding opportunities are available for the following and other sustainable energy systems:

1. Solar photo voltaic—roof or building integrated
2. Solar thermal—domestic hot water heating or building heating
3. Wind—on-site urban wind turbines
4. Geothermal, lake, river, or sea cooling

5. Cogeneration—turbines, reciprocating engines, or microturbines
6. High-performance glazing—low E coating
7. Thermal break curtain wall systems
8. Thermal storage
9. Overall building performance
10. Daylight harvesting and dimming
11. High-performance mechanical equipment such as chillers and variable frequency drives
12. Electrical systems—power factor reduction equipment

CHAPTER 2

Radiant Cooling

HISTORY

THE HISTORY OF RADIANT SPACE HEATING IN BUILDINGS GOES BACK 2,000 YEARS. Examples are found in Korea, Syria, and Rome. All three systems use the basic concept of transferring heat from a wood-burning stove or furnace to the walls and floors of the living quarters. The path of the flue gases was routed through the underfloor plenum to the walls. In Korea, the *ondol* was primarily used for living areas. In Europe and in the Islamic world, underfloor heating was primarily used in bathhouses.

Roman hypocausts were used in public bathhouses. The hypocaust was a raised floor built on brick piles, creating an underfloor air plenum. The furnace was at one end of this plenum and the chimney at the other end, creating a pathway for flue gases. Flue gases transferred heat to the slab of the occupied space, which heated the space and the occupants through radiant heating.

The Roman hypocaust was voted by the U.K. heating and ventilation industry as the top product of all time in the Hall of Fame initiative associated with the recent H&V07 exhibition, and was featured in the December issue of *Modern Building Services Journal*.[1] In the buildings associated with the Islamic system, the hypocaust underfloor plenum were replaced with pipes of chimneys buried directly under

---

[1] *Modern Building Services Journal*, "Roman hypocaust acclaimed as top H&V product ever," April 2007.

Figure 2–1 Radiant floor cooling project, Newseum, Washington, D.C. *Asif Syed*

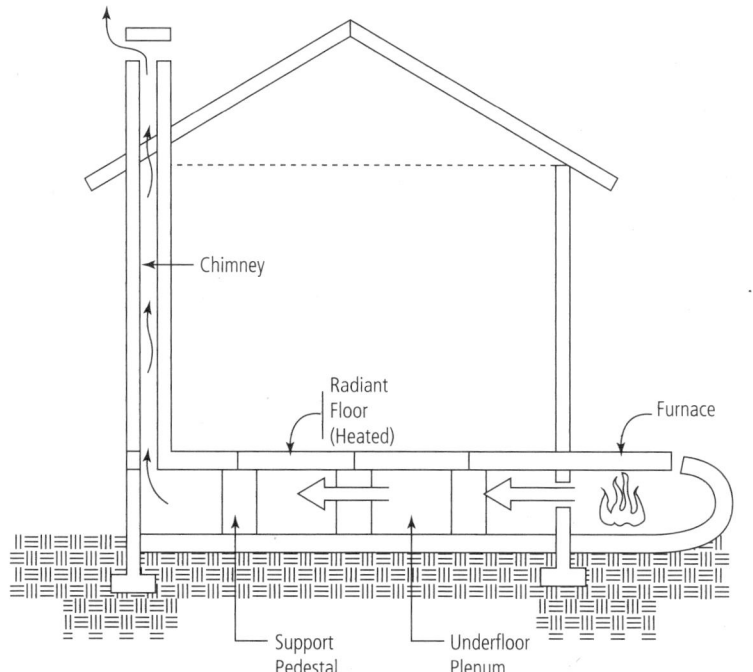

Figure 2–2 Roman hypocaust. *Asif Syed*

the floor.[2] Radiant heat is common in North America in residential construction and single-family homes. It is still popular and considered a better-quality heating method than hot air furnace and fin tube radiators.

INTRODUCTION

Life on Earth continues to exist and thrive on radiant energy from the sun. The radiation from the sun is the primary driver for the ecosystem. Plants rely on radiation for photosynthesis and in turn produce food. The Earth's surface temperature is maintained from the cycle of radiation of heat inward during the day and outward during the night. There is no medium such as air to transfer this heat. The radiant energy travels in electromagnetic waves, some of which are visible to our eye and others are not visible. Sunbathing in cold weather is a good example of radiant heating. The sunbather feels comfortable at a colder temperature due to the radiation from the sun. Similarly, in summer, when the outdoor air temperature is high at 90°F, one standing under the shade of a tree feels cooler than the temperature of the air. The leaves of the tree act as the sink for the heat from the body. The radiant method of transferring heat does not require a heat transfer medium, such as air. The energy efficiency comes from eliminating the fan system to transfer the air medium. The hot or cold object that is at a distance away radiates energy directly to our skin, which is a sense organ. Skin temperature greatly affects our comfort level. The human body maintains a steady core temperature of 98.6°F. At any increase or decrease in the core temperature, the first human organ to respond is the skin. When the core temperature increases, the blood flow to the skin increases, causing sweat, which evaporates, drawing energy from the skin and cooling the body. When the core temperature falls, the first response is reduction of blood flow to the skin, and the second response is internal heat production. A comfortable state for a human being is the state at which there is neither the need to cool the skin nor the need to warm the skin. Thermal comfort is defined by International Standards Organization Standard ISO 7730 as "the state of mind that is satisfied with the surrounding environment." To measure the comfort level in the space, the most effective tool is to measure the variables that trigger discomfort and the body's response to it: the loss of heat or the need to be cooler.

Human thermal comfort is a function of six variables[3]:

- The air temperature in the environment
- The air velocity in the environment
- The mean radiant temperature

---

[2] Peterson, Andrew. *Dictionary of Islamic Architecture*. Routledge, New York: 1996.
[3] "Comfort with DOAS Radiant Cooling System," S. A. Mumma, *ASHRAE Journal*, 2004.

- The humidity in the environment
- Clothing
- The person's metabolic rate

When clothing and metabolic rate are made constants, the mean radiant temperature and air temperature are the most influential parameters in establishing the comfort level. The mean radiant temperature interacts directly with the skin, which is the first responding human organ when discomfort is encountered. Mean radiant temperature is a complex parameter to measure; it is defined as the equivalent temperature at which the body in an environment would lose heat, if the surrounding surfaces were a matte black. A new measurement that combines the air temperature and mean radiant temperature is the operative temperature.[4] At a low velocity of air in the environment, operative temperature is an average of air and mean radiant temperature. This makes the radiant temperature an important factor in determining the comfort level of an environment.

Radiant cooling is generally a water-based system. However, there are technologies where hot or cold air is circulated through hollow concrete slabs, which are heated or cooled and radiate energy out. In the case of hollow concrete slabs, the slabs also act as a medium to store heat, leading to thermal energy storage. In a radiant cooling system the radiant surface, or slab, generally does not store heat. In a water-based system, water is the medium that is used to transfer heat from the space to the outdoors, where it is rejected. Because water has a higher density and carries more heat than air, a water-based system is far more efficient than an all-air system. Pumps use far less energy than fans because they operate at higher efficiencies than fans. The standard fan efficiency is 65 percent, whereas the standard pump efficiency is 85 percent. The heat-carrying capacity and the fan equipment efficiency are the major factors in the lower energy costs of the radiant cooling system.

There are three common technologies available:

1. Radiant panels: Radiant ceiling panels look very similar to acoustic tiles, and some have the same finish as acoustic tiles. The panel is made up of aluminum, with copper tubes on the back side of the panel. These are the most common type of radiant ceiling, commonly used in office buildings, hospitals, and schools.

2. Radiant floors: Radiant floors are gaining popularity throughout the United States, and have been part of many high-profile projects. Polyethylene tubes are embedded in the structural slab and covered with concrete. These installations are suitable for large public spaces and residential applications,

---

[4] "Radiant Floor Cooling Systems," Bjarne Olesen, *ASHRAE Journal*, September 2008.

including college dormitories, hotels, single-family homes, and apartment buildings.

3. Chilled walls or ceilings: In this system, the capillary tubes are embedded in plaster on walls and ceilings. Though commonly referred to as chilled ceilings, these systems can also be used for heating, by circulating hot water. These systems are also known as capillary tube systems.

Radiant heating is a popular home heating system that uses water tubes embedded in the floor construction. Owners of homes with radiant heating are very satisfied with the comfort level, the lower heating bills, and the quiet operation. The thermostat in a home with radiant heating is normally set at 65°F, producing the same level of comfort as an air-heated system, which is typically set much higher, at 72 to 75°F. The boiler operating temperatures are much lower in a radiant system, 100 to 120°F, compared to 180 to 200°F in an air system. The lower setpoint leads to lower consumption of fossil fuel and lowers energy bills. Homes with radiant heating systems have higher market value because the home real estate market correctly considers this the superior heating system. Hot water from a boiler is circulated through the tubes, which heat the floors. The floors radiate heat to the occupants, walls, furnishings, and ceilings. In the 1970s, copper tubes were used, but these were corrosive with concrete. Copper is not used anymore in the radiant heating of homes; the current technology for tubes is polyethylene (PEX). These tubes are noncorrosive and have sufficient conductivity to transfer heat to the floor.

Figure 2–3 PEX tubing.
Picture by J. Macaluso

WHY RADIANT COOLING?

Radiant cooling has several advantages over conventional systems. The most common conventional system is a variable air volume (VAV) air distribution. The benefits of radiant cooling systems are:

1. Offers better comfort—direct interaction with skin
2. Smaller ducts in the space—higher ceilings and lower costs
3. Smaller fan systems—smaller mechanical systems and lower costs
4. Energy efficient—sustainability and lower energy costs

In a radiant cooling system, the circulation of air is limited to ventilating the space to allow fresh outdoor air and oxygen to enter the space and to remove any contaminants generated in the space. In most applications, the ventilating air is only a fraction of the air required for cooling. In a standard office space, about 20 percent of the air circulated is for ventilation, and 80 percent is for cooling. Radiant cooling provides an opportunity to reduce the amount of air required for cooling. Heat loads in the space are cooled directly by radiation, eliminating or reducing air circulation. The reduction in air circulation results in lower airflow, leading to smaller ducts and fans. The chilled water operating temperature of radiant floors is higher than that of conventional systems. The chilled water temperature of the radiant system is about 55°F, whereas the conventional systems use 45°F water. The higher chilling temperature lends itself to good use of geothermal energy. The subsurface temperature of the Earth, from which geothermal energy is extracted, is a constant 55°F, which works well for geothermal heat pumps.

Modern design practices and control strategies completely eliminate condensation. Condensation has been a concern when it comes to radiant cooling. The cool surfaces provide an opportunity for condensation from the air in the room. Condensation happens when the temperature of the radiant surface drops below the dew point of the air in the space. Some early designs in the 1970s got bad press because of condensation, but this was primarily due to incorrect operation of the system and use of standard chilled water (colder) in the radiant systems. Modern designs use a higher temperature of chilled water, avoiding condensation completely. The temperature of the water circulating to the radiant surfaces is selected by design to be higher than the dew point of the air in the space. Additionally, controls are set up to constantly monitor dew point temperature in the space and maintain the chilled water temperature above the dew point. Another control strategy to eliminate condensation is window sensors; when windows are open, the control system blocks the chilled water from the radiant panel. When the outdoors is humid and windows are left open, the humidity levels can increase the dew point. The window sensors are similar to security alarm sensors, and their cost is not prohibitive.

Energy savings from radiant cooling systems come from:

1. The system uses a higher chilled water temperature of 55°F (45°F in a conventional system), which means the compressors have to work less, using less energy.
2. The use of radiant cooling systems reduces airflow rates, by about 50 to 60 percent. This allows for smaller fans and motors, which consume less energy to circulate air.
3. Smaller fans generate less heat. Fans are generally very inefficient devices. Most fans in building systems operate at about 65 percent efficiency. About 35 percent of the energy that goes into driving the fan is converted into heat, heating the air that is meant to cool the spaces.
4. Space temperatures can be maintained slightly higher than the conventional systems.
5. Energy savings from radiant cooling systems can average about 30 percent.[5]

TABLE 2–1 THE DIFFERENCES BETWEEN RADIANT SYSTEMS AND CONVENTIONAL SYSTEMS

No.	Conventional System (Air Distribution)	Radiant System
1	Sole method of heat extraction from heat-generating objects is by circulation of air around the objects.	Primary method of heat extraction is by radiation between the cool radiant surface and the heat-generating object. Secondary method is by air circulation.
2	Higher quantity of air circulation.	Air quantity is reduced as a portion of heat is transferred by radiation, which is carried by water.
3	Higher noise from diffusers and ducts due to high velocity. Higher noise is due to more air circulation. Larger ducts to carry the air, which occupy ceiling space.	Lower noise levels from smaller ducts and lower velocities. Small air quantity circulation has less noise. Smaller ducts allow increased ceiling heights.
4	Larger fan and air-handling equipment with larger mechanical rooms.	Smaller fan and air-handling units reduce the size of mechanical rooms.
5	Higher fan energy to circulate the air.	Pump energy to transfer heat is less, compared to fan energy.
6	Coordination with architect is limited to location of diffusers and ducts in the ceilings.	Additional coordination is required to locate the radiant surfaces, such as ceiling panels, wall panels, and pipes in floors for radiant floors.
7	Condensation is not a concern.	Condensation is a concern, but can be avoided by proper design and advanced controls. The advanced controls include continuous monitoring of dew point of air in the space and making adjustments to maintain the water temperature above the dew point.
8	Response time for change in temperature setting of the space is quick. System operation can be changed from heating to cooling or vice versa with quick response.	Response time to temperature settings of the space is slow. Change from heating to cooling must be carefully planned ahead of time, using weather forecasts.
9	Chilled water temperature used is industry standard of 45°F.	Chilled water temperature used is above the dew point temperature of the space. Normally 55°F water is used and is adjusted based on indoor humidity conditions.

---

[5] "Energy and peak power savings potential of radiant cooling systems in US commercial buildings," Corina Stetiu (Lawrence Berkeley National Laboratory, Berkeley, CA), *Energy and Buildings* 30(2): 1999.

APPLICATIONS

Radiant cooling can be used in a variety of new buildings, and in existing buildings to increase the capacity of the existing systems. The typical applications are:

1. Public spaces
2. Office buildings
3. Laboratories
4. Residential buildings
5. Hospitals
6. Historic preservation

PUBLIC SPACES, RADIANT COOLING FLOORS

Radiant cooling is becoming common in public spaces such as large gathering spaces, building lobbies, airport terminals, cafeterias, train stations, and the like. The radiant cooling floor is the most common approach in these spaces. Recent high-profile projects include the Bangkok International Airport; the Becton, Dickinson and Company headquarters cafeteria in Franklin Lakes, NJ; the Newseum in Washington, D.C.; and the Hearst headquarters lobby in New York City. The advanced technology of radiant cooling is relatively new in these geographical areas, but it has proven to be successful. The building owners, operators, and occupants have a very high satisfaction with radiant cooling systems. Radiant cooling floors have the following benefits in large public spaces:

1. The opportunity for the space to have large skylights. Heat load from the skylights (solar radiation) is absorbed directly into the floor. Heat is transferred out of the space with water, a far more efficient way than air.

2. High ceilings or open spaces provide opportunity for stratification and spot conditioning. Stratification allows buildup of warm air above the occupied zone. This avoids having to cool and heat spaces or volumes of space above the occupant height.

3. Bigger skylights can be used. Energy codes limit the skylight area, unless a performance approach demonstrates that the energy used is less than that in a conventional system. The energy efficiency of the radiant cooling system permits the trade-off with larger skylights.

4. Less installation of large ducts and grilles in the space, which are generally obtrusive. The biggest architectural benefit is that the heating system is integrated with the floor, minimizing visible grilles and ductwork.

5. The large area of the floor provides the best angle of incidence for radiance to the occupant. The angle between the average radiant area and the occupant

surface produces the best mean radiant temperature, leading to effective thermal comfort.

6. Multiple zones of radiant floor can be created, providing efficient zone controls and better energy efficiency.

Figure 2–4a Radiant cooling floor Newseum, Washington, D.C. *Asif Syed*

Figure 2–4b Newseum radiant cooling floor walkway bridge. *Asif Syed*

Figure 2–5a Becton, Dickinson and Company cafeteria. *Photograph © Brad Feinknopf 2008*

Figure 2–5b Radiant pipe manifold. *Photo taken and provided by BHC Architects*

FLOOR CONSTRUCTION

A series of tubes is installed under the floor. The tubes are made from polyethylene (PEX), which is noncorrosive. The floor construction increases by 1½ to 1¾ inches. The construction of the floor includes ½-inch insulation above the structural slab, which prevents heat transfer to the structural slab and to the plenum below. The insulation

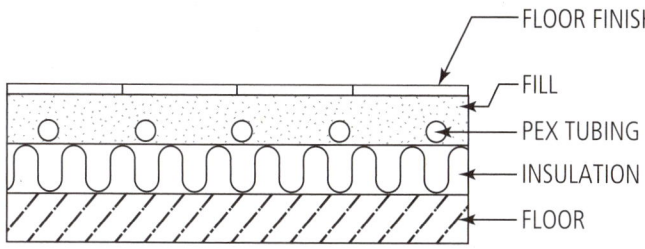

Figure 2–6 Floor sandwich of radiant cooling floor. *Asif Syed*

material is rigid foam insulation. The polyethylene tubes are laid on top of the insulation and stapled to the insulation. The pipes can range in size from ½ to 1 inch, based on the cooling capacity of the design. Normal practice is ½- and ⅝-inch pipes. A concrete fill is installed to cover the tubes. The total thickness of the fill above the insulation is 1¼ to 1½ inches. On top of the concrete fill is the floor-covering material. Ideal floor-covering materials are those that are good conductors of heat. Common materials for radiant cooling floors are tile, stone, or other materials that are good conductors of heat. Materials that are insulators must be avoided or carefully evaluated for reduction in performance. Carpet is not recommended and normally not used. Densely padded thin carpet and wood are used but are not recommended, because of the low heat conductivity of the materials.

OFFICE BUILDINGS, RADIANT CEILING PANELS

Radiant cooling in the office environment is provided through panels in the ceilings. Most office buildings have ceilings that provide an excellent opportunity to use radiant cooling. The cooling loads in office buildings have generally been reduced. Standard practice in the 1980s was 5 watts per square foot. This number has now been reduced as a result of more efficient office equipment and lighting. Lighting heat load gain has gone down from 1.5 to 1 watt per square foot. Similarly, thanks to LCD/LED screens, office plug load power has gone down from 3.5 to 2.5 watts per square foot. These heat loads are indicative of most office buildings; however, they can vary based on the occupant density and activity. Light trading or extensive use of computers can have higher loads. A detailed analysis of heat loads is required in the design and selection of radiant ceiling panels. This reduction increases the potential to use radiant ceilings. The lower loads mean that the radiant cooling panels can cool a higher percentage of the cooling load. In existing office buildings where the capacity has to be increased, radiant ceilings can be installed without any addition of ductwork and fans, both of which are bulky and voluminous, making them difficult to install. It should be noted that radiant ceiling systems work with ducted air distribution. The air distribution is about 50 to 60 percent less, and is for ventilation and moisture removal. The ceiling panels are available in standard ceiling tile patterns or other finishes. Benefits of radiant cooling systems are:

1. A draft-free environment is achieved. Drafts are a main source of complaints from occupants.
2. A lower noise level is achieved. Reduced airflow reduces duct- and equipment-generated noise.
3. The system does not take up any floor space, but rather gives back floor space in the building because of reduced airshafts and smaller mechanical rooms.
4. Existing buildings can be retrofitted with radiant panels. In rooms with lower floor-to-floor heights, where large ducts cannot be accommodated, radiant panels can be used.
5. The system offers energy savings and lower operating costs.
6. In new buildings, floor-to-floor heights can be reduced. This offers a big advantage in high-rise buildings. A reduction of 6 inches per floor can result in two additional floors in a fifty-floor building.

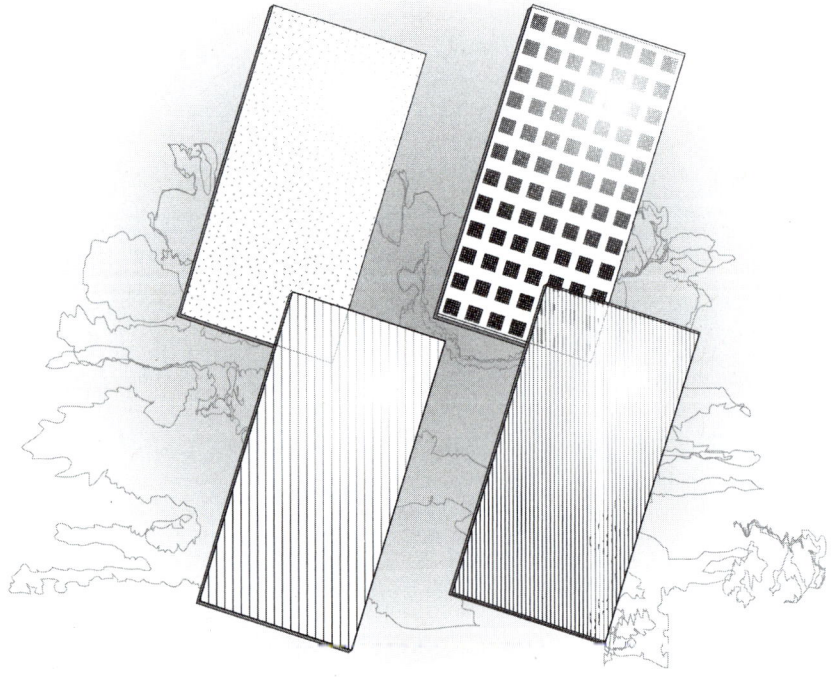

Figure 2–7 Radiant ceiling panels in an office building. *Aero Tech Mfg. Inc.*

Figure 2–8 Radiant ceiling panel finishes. *Aero Tech Mfg. Inc.*

Office spaces in the building are of two basic types: interior spaces, which do not have exterior or envelope loads, and perimeter spaces, which have solar and conduction loads from the envelope. The load in the interior spaces is made up of office equipment, lights, and occupants. The capacity of the radiant ceiling panels may generally meet the sensible loads of the interior space. The ceiling area required can be in the range of 30 to 50 percent, based on the interior loads. For the perimeter office spaces, in addition to interior loads there are envelope loads of solar heat gain and heat conduction. It is important that the envelope loads be reduced with high-performance glass, shading, and insulation. In most buildings, all the sensible loads of the perimeter cannot be met by radiant cooling panels; additional cooling is required. However, radiant cooling panels are effective at the perimeter because they efficiently cool the surfaces that are heated from solar heat gains. Perimeter heat loads increase the peak demand for cooling, increasing the size of the air circulation system for only a few hours in the day. Radiant cooling panels along the perimeter can offset the peak demand, reducing the size of the air-handling units. Radiant cooling panels are available in the same sizes and configurations as standard ceiling tiles. The finishes of the panels can be made to match the ceiling tiles.

LABORATORIES: RADIANT CEILING PANELS

Laboratories are very high consumers of energy and are considered the gas guzzlers of buildings. This is primarily due to once-through air systems. For reasons of safety, a once-through air system with a high air change rate is required. Though energy codes mandate that the heat or energy from exhaust air streams be recovered, in real life the systems do not work well and the efficiency of heat recovery is very low. The best opportunity to reduce the air flow is by managing the high internal loads. Radiant ceiling panels in the lab spaces can absorb the heat generated by equipment, reducing the air flow. Active chilled beams are more popular in labs, but radiant panels to be used in new lab buildings and retrofit of older laboratories. Chilled beams will be discussed in a separate chapter.

Most older labs built in 60s, 70s, and 80s are in the process of retrofitting to save energy. The retrofits are mostly driven by energy efficiency and to reduce the operating cost. Some renovations are geared toward making the labs flexible for use in different types of research and teaching. The most common retrofit is to save energy by lowering the air flow through the fume hoods. This reduces the overall air quantity and lowers the heating and cooling costs. This works well for labs with high fume hood density. For laboratories that are not hood-intensive, radiant ceiling panels present an excellent opportunity to lower the air change rate. This saves energy because of not having to cool the air that was eliminated by radiant cooling.

Flexibility is the buzzword in design for labs. Most new labs need the flexibility to accommodate a variety of research types, with varying heat loads. The average research project can vary in length from six months to a few years. Installing infrastructure for the maximum predicted heat load has been the common practice in the past, and it continues to date. This is wasteful of energy, especially in these energy-conscious times. There are no codes that specify the proper infrastructure. It is up to the discretion of the designers and owners to decide what is appropriate. Once the air system infrastructure is installed, there is very limited opportunity for turndown, or none at all. A modular radiant cooling panel system can assist in optimizing and reducing the size of the air system's infrastructure. Radiant panels can be installed and removed as needed; or just capped off, or moved to labs with higher heat loads, providing a flexible system. The advantages of radiant ceiling panels in laboratory buildings include:

1. The biggest benefit of radiant cooling is the reduction in operating costs, which are about four times those of a standard office building. This reduces energy consumption and fossil fuel burning.

2. Reduction in air flow reduces 24/7 and 365-day cooling or heating.

3. Reduction in air flow reduces fan power.

4. Radiant cooling provides a flexible system that can be moved, added to, removed, and isolated to meet the varying needs of research and lab heat load configuration.

RESIDENTIAL

Radiant cooling and heating can be used in all types of residential buildings such as college dormitories, large multiple-family high-rise buildings, single-family homes, and hotels. In all applications the energy reduction is significant, about 42 percent in a college dormitory building.[6] The savings are similar in other residential applications. In all residential systems, floors and ceilings are used as the primary radiant surface. Residential radiant flooring has an extra challenge, as carpets are a common residential floor finish. Carpets, which are insulating material, reduce the heat transfer from radiation. It becomes important to select floor materials that are not insulating, such as pressed wood, ceramic tile, stone, or any material with moderate to good thermal conductivity. Even with good thermally conductive material finish on the floor, residential units have rugs in the living spaces. Ceiling and walls can also be used as radiant surfaces. The installation can vary based on the number of zone controls. It is generally recommended to have separate zones for living areas and bedrooms, to take advantage of the diversity. Both bedrooms and living areas are rarely occupied at the same time, allowing temperature setback during unoccupied or non-use time. This feature can significantly save energy. Most common conventional residential systems are packaged terminal air conditioner (PTAC), heat pump, and fan coil units. In most buildings, for a conventional system, a central air system supplies air for ventilation, or provides make-up air, and exhausts air from the toilets and kitchen. Most codes permit the use of operable windows as a source of ventilation. However, it is a common practice in current designs to have a central make-up air system and usually the make-up air is conditioned (heated and or cooled and dehumidified). This central make-up air system remains almost identical in a radiant system. The make-up air which provides ventilation also removes moisture from the space. Almost all residential radiant heating and cooling systems are two-pipe systems, in which the same set of pipes provide both heating and cooling, with seasonal switching between hot and cold water. The response time during switchover from heating to cooling and vice versa is slow. This usually happens in seasons when sudden swings occur in outdoor temperature. The response time is slower because the system has to heat or cool the mass associated with the system such as floors or slabs or walls, where the tubes are,

---

[6] Dartmouth Sustainability Initiative, "Sustainable High Performance Buildings," www.dartmouth.edu/~sustain/dartmouth/building.html.

before it radiates to the surroundings. An educated and informed user can generally deal with this issue based on pre-programming based on weather forecasts. The radiant cooling control system normally has an outdoor temperature sensor, but there are no protocols or software packages that can integrate weather forecasts from weather information websites.

COLLEGE DORMITORIES

College dormitories are similar to multifamily residences. The main difference is that dormitories are communal living and there is normally some relaxation on the extent of zones of temperature control. It is possible to have a common temperature setpoint for multiple apartments and also apartments on upper and lower floors. This leads to a different approach to radiant cooling. Common construction in college dormitories is with pre-cast concrete planks or slabs with no ceilings in the bedrooms and living areas. The ceilings are limited to bathrooms, hallways, and toilets. Radiant heating and cooling works well with this type of construction. The lack of ceiling presents an opportunity to use the plank or slab for cooling and heating from the top and bottom. There are two approaches in the construction of the radiant heating and cooling tubing: multidirectional and single direction. Multidirectional radiation provides radiation above the slab and below the slab. Multidirectional systems are required for carpeted floors. Multidirectional radiation is achieved by eliminating the insulation between the pipe and concrete plank or slab. Multidirectional slabs provide a communal comfort zone—that is, one common thermostat will control temperature for apartment units on both floors. This arrangement is generally acceptable in dormitories; temperature ranges can be maintained within acceptable limits in both apartment units (upper and lower). Unidirectional radiation is generally sufficient when the floor is not carpeted and the rooms are large and provide enough radiant surface areas. In unidirectional systems, thermostatic zone controls are dedicated to the apartment on that floor. The ventilation systems in this design should include a make-up and exhaust fan. The make-up air provided for exhaust of toilets and kitchens normally satisfies the ventilation requirement of the occupants and removes the moisture generated in the space. In order to maximize the energy savings, the exhaust and make-up must be coupled in a heat recovery system. The required level of conditioning (heating, cooling, and dehumidification) of the make-up air is based on the internal space heat loads, moisture generation and removal, and radiant floor capacity.

In applications where floor and ceiling slabs cannot be used as radiant surfaces, then walls can be used as radiant cooling surfaces. The wall radiant cooling systems are called chilled walls. Chilled walls have capillary tubes that are only $1/16$ inch in diameter and lay behind the plaster or drywall ceilings.

APPLICATIONS 37

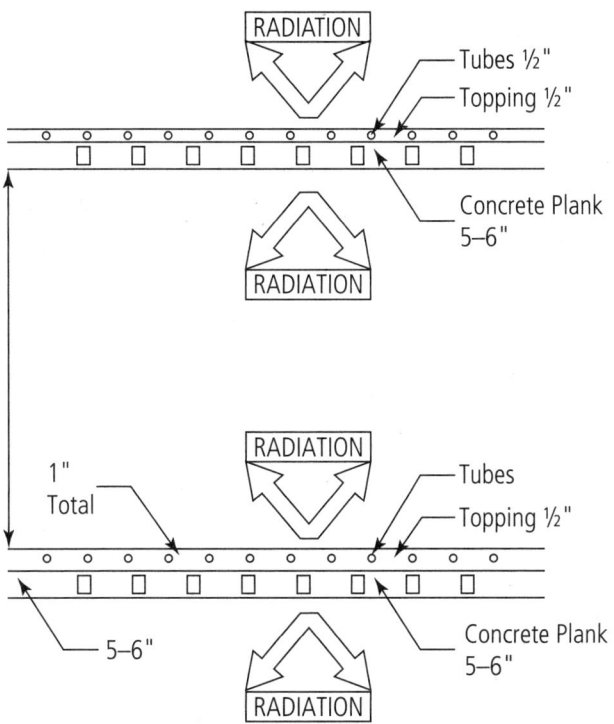

Figure 2–9 Multidirectional radiant slabs. *Asif Syed*

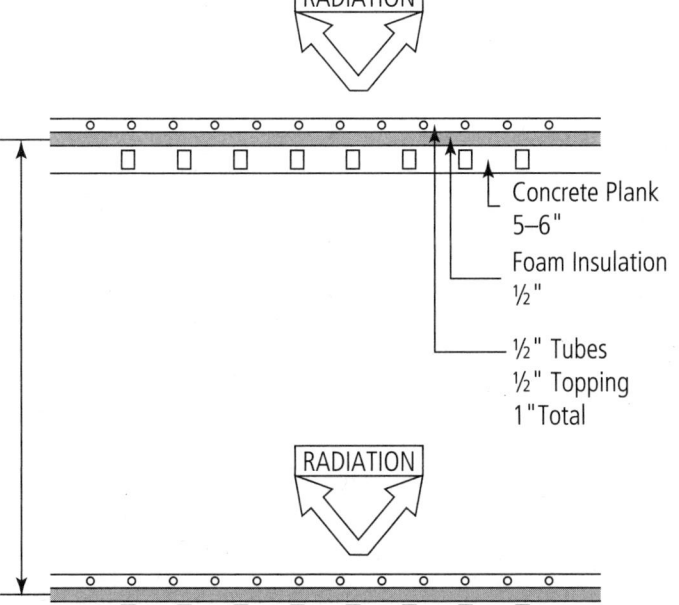

Figure 2–10 Unidirectional radiant slabs. *Asif Syed*

Figure 2–11a Capillary tubes in chilled walls and ceilings. *www.naturalcooling.com, Buenos Services Corporation*

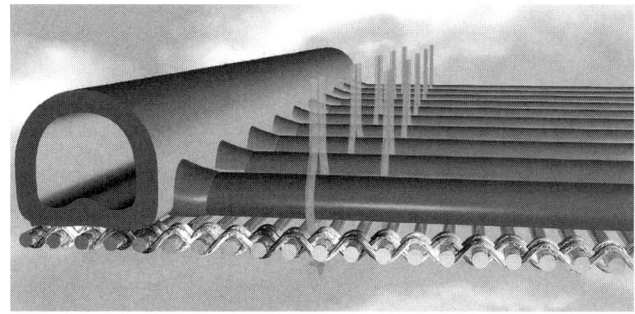

Figure 2–11b Chilled walls capillary system. *www.naturalcooling.com, Buenos Services Corporation*

HOSPITALS

Radiant cooling systems are not new to hospitals. Some hospitals built in the 1970s used radiant cooling panels or chilled ceilings in the patient rooms. The panels preformed as heaters in winter and provided cooling in summer, which was accomplished by switching the hot water to chilled water seasonally. The radiant ceiling panels were a good energy-savings feature, offered contaminant and infection control, and saved floor space. The radiant panel was in the ceiling and did not have any components to gather dust or other infectious agents. Baseboard radiators, on the other hand, took up floor space and gathered dust. The radiant cooling systems went out of fashion because of operational problems. Condensation was the main issue. The systems were relatively new to the entire building industry, and operators were not familiar with them. When the chilled water temperature went out of range, it led to condensation on the panels. As explained earlier, it is essential that all radiant cooling systems maintain chilled water temperature above the dew point. The control strategies and technology were not as advanced and sophisticated as the present-day direct digital control (DDC) systems. Today, both advanced control systems and experienced operators are available, and condensation is not an issue anymore.

With high-performance building envelopes, advanced daylighting controls, and lower internal heat gains with advanced lighting design, the cooling load in the typical hospital patient room is less than it was a few years back. This presents an opportunity to use radiant cooling systems. The minimum ventilation rate has to be maintained per codes, and any additional heat load can be offset with radiant cooling panels. The benefits of radiant cooling panels in hospitals are:

1. This system provides superior comfort, with uniform temperature, lower noise, and a draft-free environment for patients.
2. Energy-efficient, thermal energy can be carried by water-based radiant panels more effectively than by all air systems with ducts and fans.
3. The use of smaller ducts, which are dedicated just to ventilation air, relieves congestion in hospital corridor ceilings. The congestion caused by ductwork and other services leads to increase in floor-to-floor heights, increasing the cost of construction.
4. Integration into the architecture element—ceilings—eliminates the need for fan coil units and/or VAV boxes, reducing the mechanical costs.
5. The system does not gather dust or infectious agents.

RADIANT COOLING AND HISTORIC PRESERVATION

Radiant cooling presents great opportunities in historic preservation projects. Retaining the original structure, interior elements such as ceilings and walls, architectural finishes, and floor plan layouts is the primary goal of historic preservation. Buildings built 80 to 100 years ago were not designed for air conditioning and had only heating components: steam radiators or fireplaces. Adding air conditioning with conventional all-air duct systems—with large ducts, VAV boxes, air shafts, and large fan rooms—makes it almost impossible to retain the original character of the building. Ducts and VAV boxes require close to 2 feet of plenum space, which is not available in historic buildings. Moving air from roofs and mechanical rooms requires shafts, which are also not available. Radiant heating and cooling can be conveniently integrated with the original architecture of these buildings. The University of Michigan's Dana Building, a historic landmark, uses radiant ceiling cooling panels and cuts energy costs by 20 percent, compared to the baseline all-air system.[7]

---

[7] LEED Energy Performance Modeling and Evaluation of the S.T. Dana Building Renovations, Sharada Gundala, master's thesis, University of Michigan: Ann Arbor, 2003.

CHAPTER 3

Displacement Ventilation

HISTORY

THE HISTORY OF DISPLACEMENT VENTILATION DATES BACK TO THE EARLY YEARS OF industrial ventilation.[1] Displacement ventilation kept the factory floors flooded with fresh air for workers, while keeping the contaminants out. The primary goal of displacement ventilation was and is to provide fresh or outside air that is not mixed with contaminants. In the 1970s, displacement was used for contaminant control in factories,[2] to protect the health and welfare of workers. The protection from contaminants was noticeable, and workers in polluted environments were provided with fresh air. Early examples of displacement ventilation can be found in theaters, where air was supplied from beneath the seats. Carnegie Hall, built in 1891, had its air supply under the seats. In the 1980s, displacement ventilation for commercial building spaces was first developed and adopted in Scandinavia. From the 1990s onward, it has gained popularity in Europe. With the growing importance of sustainability and energy efficiency, it is beginning to gain worldwide popularity. Its benefits of energy efficiency combined with superior comfort and improved indoor air quality are the primary reasons for its success. It is still relatively new in the United States, but is gaining popularity.

Why is displacement ventilation not used in all the buildings as a conventional system? This is a very valid question. Our current air conditioning approach started

---

[1] W. Baturin, *Fundamentals of Industrial Ventilation*, NY: Pergamon Press, 1972.
[2] Federation of European Heating, Ventilating and Air-conditioning Associations (REHVA), *Displacement Ventilation in Non-industrial Premises*, 2002.

with retrofitting existing buildings with air conditioning. As air conditioning became popular for comfort, it was fitted into existing buildings. It was just easier to install ductwork into the head room or ceiling space of existing buildings to distribute air, rather than to use the floor space and disrupting the carpet area. If the air conditioning and building had evolved together, then displacement would have become the conventional system.

INTRODUCTION

Displacement ventilation systems help in solving the biggest problem we face today: energy. Displacement ventilation is inherently an energy-savings system. In addition, it helps to maintain a better indoor environment, with increased comfort and lower contaminant levels, contributing toward the overall sustainability of the building.

Displacement ventilation is a method of ventilating the space by supplying air at floor level and returning air at high or ceiling level. It is the opposite of the conventional air conditioning methods, whereby air is supplied from the ceiling and also returned at the ceiling. The process of displacement utilizes the force of buoyancy. Buoyancy is a natural force that makes lighter or low-density mass rise or float, especially fluids or air. When heat from an object such as office equipment or a human body is released, it increases the temperature of the air around it, which makes the air lighter or less dense to go up, or displace itself to a higher elevation, until it is returned from a ceiling grille. Fresh new air from the floor rises to replace the displaced air.

In displacement ventilation systems, the supply air temperature is higher than that of the conventional systems by 10°F. The supply air temperature of a displacement ventilation system is 65°F (55°F for conventional systems). The return air temperature is also higher than that of the conventional air system by 10°F. The return air temperature of a displacement ventilation system is 85°F (75°F in conventional systems). In the displacement system, cold air supplied at floor level moves sideways or horizontally, "displacing" the air being moved up by the natural buoyancy created by heat sources. As the displaced (warm) air moves up, it gains more heat from the lights, and at the ceiling the temperature of the air is 85°F. This warm air is then returned to the cooling apparatus. The velocity of air discharge at floor level is low, so there are no drafts. The temperature of the air is high enough (65°F) that it does not feel cold.

CONVENTIONAL OR MIXED-AIR SYSTEMS

The working of a displacement system is best understood by comparing it to the conventional air distribution system. The most common air distribution system is known as a mixed-air system. The cold air supplied from the ceiling thoroughly churns all the air in the room to create a uniform temperature. The warm air in the room, which is found around heat-generating objects, is "mixed" with cold air to form a uniform temperature throughout the room—therefore, the name "mixed air system." Its drawbacks are higher energy consumption and a higher concentration of contaminants. Because most applications, such as air conditioning, are for human comfort, the entire space volume does not have to be conditioned, just the occupied zone of 6 feet. Cooling the nonoccupied volume of air above the occupied zone is a waste of energy. The mixing of air also mixes in contaminants, increasing their concentration. In the mixed-air system, air is normally supplied at 55°F, and the space is maintained at 75°F. The air from the room is returned to the cooling apparatus (air-handling units or air conditioning units) at 75°F.

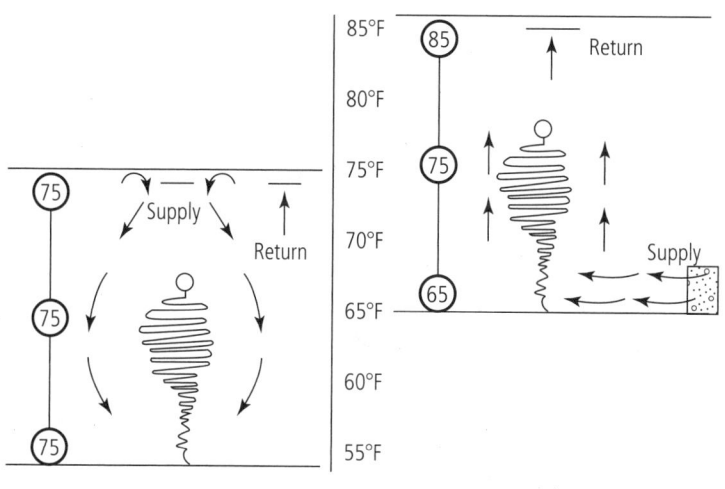

Figure 3–1 The principle of displacement. *Asif Syed*

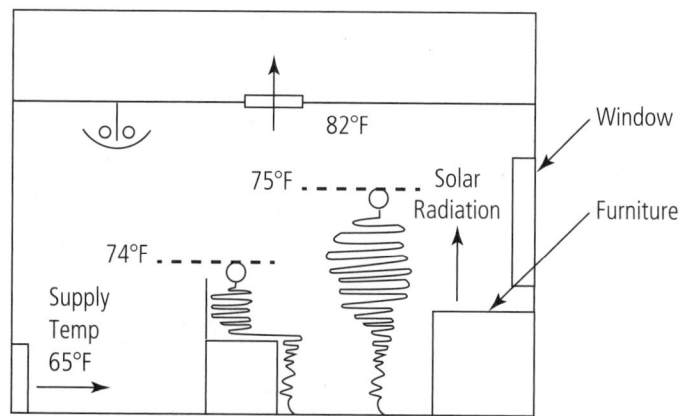

Figure 3–2 Displacement temperature gradient. *Asif Syed*

Figure 3–3 Trox displacement flow diffusers. *TROX Technik*

Benefits of displacement ventilation:

1. Energy savings: Reduced energy consumption in displacement ventilation comes from:

    a. Higher air temperature: The supply air temperature of the displacement system is higher than that of a conventional overhead air distribution system. The UFAD supply air temperature is 65°F, whereas conventional air systems use 55°F air. The benefit from the higher temperature is twofold:

    i. The cooling system or the compressors will not have to work as hard and will therefore use less energy. The refrigeration process of absorbing heat from a high-temperature source (indoors) and rejecting it to the high-temperature sink (outdoors) is more efficient because the difference between source and sink is less.

    ii. The economizer, or free cooling, control method is used in air conditioning systems. This method uses the outdoor ambient cooler temperature to cool the indoor heat-generating space. Office interior spaces require cooling even in the winter, when outdoor temperatures are cold. For displacement systems, the economizer period can be extended. In conventional systems, economizer or free cooling stops when the outdoor temperature reaches 55°F. In displacement systems, the free cooling can continue up to 65°F.

b. Thermal stratification: Displacement systems stratify (not condition) the space above head level up to the underside of the ceiling with a higher temperature (80 to 85°F). From the floor to just above head level is considered the occupied zone and is maintained at the comfort temperature of 72 to 78°F. Conventional systems maintain a uniform comfort temperature (72 to 78°F) throughout the entire height of the occupied space. In spaces with high ceilings, the occupant zone is about 6 feet high. As air passes through this occupied zone, its temperature is suitable for human comfort (72 to 78°F). Air becomes warmer after passing the occupied zone (usually 6 feet from floor), but at that point it does not matter. In the displacement ventilation system, the comfort zone or volume of space air conditioned is limited to the occupied zone. In a space with a 10-foot ceiling, the occupied space is 6 feet. So only about 60 percent of the volume of the space is air conditioned to the human comfort level, and the remaining 40 percent, above the human comfort zone, is unconditioned. This percentage of unconditioned space in nonoccupied zones increases as the ceiling height increases, further increasing the savings in energy needed to operate the system.

2. Indoor environment:

The indoor environment is improved by contaminant control and ventilation effectiveness in displacement systems. Both benefits are due to the physics of the operating principle of displacement systems. Conventional air distribution systems, also known as mixed-air systems, lack these benefits because of their air delivery methodology.

a. Contaminant levels: Contaminants generated at the floor level, from copy machines or occupants, directly rise along with the displaced air to the ceiling and into the return air grilles. Displacement systems limit the churning movement of the air, reducing the volumetric space for contaminants and decreasing their density or parts per million (PPM) count.

b. Ventilation effectiveness: This feature is defined as the effectiveness of an air distribution system in removing internally generated pollutants from the ventilated space. Displacement systems are more effective than conventional systems and carry a higher rating.

   c. Noise levels: The noise levels of displacement systems are low, as a result of the lower velocity through air outlets. The velocity of a displacement air outlet is about five times slower than that of a conventional system. High air velocity is the major source of noise in conventional overhead systems.

   d. Thermal comfort: Hot and cold spots are the top complaints in buildings. The complaints are due to drafts and inadequate thermostats or zones of control. Both drafts and inadequate zones are inherent to conventional overhead design.

3. Simple and easy: It is middle-school science. Middle-school students, who are not influenced by our design methods (or have not been corrupted by our bad practices in design), can understand the basic principles of physics that give the displacement system its benefits of comfort, quiet, and contaminant control.

TABLE 3–1 THE DIFFERENCES BETWEEN CONVENTIONAL AND DISPLACEMENT VENTILATION SYSTEMS, INCLUDING BENEFITS

No.	Displacement	Conventional
1	Supply air temperature is 65°F and return air temperature is 85°F.	Supply air temperature is 55°F and return air temperature is 75°F.
2	Supply is at floor level.	Supply is at ceiling elevation.
3	Air outlets are floor grilles, low side wall grilles, and manufactured units. See Figure 3–3 for types of air outlets available.	Air outlets are standard ceiling diffusers, which are louver face, slots, or high side wall grilles.
4	Supply air velocity through the floor air outlet is 50 to 70 feet per minute.	Supply air velocity through ceiling diffusers is 500 feet per minute.
5	Larger size air outlet is required to maintain low velocity. Air outlet or grille sizes are about 7 to 10 times larger. Here architect and engineer have to work together closely to locate the air outlets.	Standard ceiling diffusers are required. Level of coordination between architect and engineer is less.
6	Lower noise level results from low air velocity through the air outlet.	Higher noise levels from higher air velocity through the outlet. Diffuser noise is one of the main sources of noise in the HVAC system.
7	Total air quantity flowing through the space is the same as in conventional systems.	Total air quantity flowing through the space is the same as in displacement systems.
8	Quantity of ventilation or outside or fresh air can be reduced to provide the equivalent level of indoor air quality as a conventional system. This reduces energy consumption as outside air is hot and humid (summer) or cold and dry (winter) and energy is used to condition it to room comfort level.	For the same indoor air quality, higher amounts of outside or ventilation or fresh air are required. This increases the energy costs to cool and heat the additional air quantity.

TABLE 3-1 (CONTINUED)

No.	Displacement	Conventional
9	Cooling apparatus (air-handling unit) is different from conventional. To cool and dehumidify air to 65°F, only a desired fraction of air is dehumidified by cooling to 55°F, and then mixed with warmer room air to create 65°F air. This is an energy-efficient process, as dehumidifying air uses a large amount of the cooling energy.	Cooling apparatus is standard. Conventional units cool and dehumidify air to 55°F. Here the entire volume of air is dehumidified. In most applications, the extent of dehumidification is more than needed, but the system requires it to cool the air to maintain the desired temperature. This is an inefficient and energy-consuming process.
10	In a free cooling cycle, outside air up to 65°F can be used, giving more hours of free cooling. Free cooling is defined as a means of using the outside environment to cool the indoors (with acceptable humidity).	In a free cooling cycle, only outside air up to 55°F can be used, reducing the number of free cooling hours (with acceptable humidity).
11	In displacement ventilation, the air is separated into two streams: one for dehumidification and cooling, and another for reheating. This saves energy by not overcooling or over-dehumidifying.	Entire supply air is cooled and dehumidified in one single process. This results in overcooling or over-dehumidification of air, a waste of energy.
12	Contaminant level (PPM) is lower. The displacement principle causes air from the contaminant source to rise directly into the return grille.	Higher PPM of contaminants. Mixing of air in the entire room causes contaminants to increase in concentration.
13	No short circuit between supply and exhaust, providing extremely effective ventilation of the space. Supply and extract are located at opposite ends and also at different elevations.	Good possibility of short circuit between supply and exhaust, reducing the effectiveness of ventilation. Supply and extract are located at the same elevation.
14	Low velocity and the displacement principle prevent cold drafts.	Usually drafty environment or dumping of cold air due to reduced flow.
15	Uses less energy.	Uses more energy.

DIFFERENCE BETWEEN DISPLACEMENT AND UNDERFLOOR AIR DISTRIBUTION (UFAD)

These two systems are frequently mistaken as one and the same. They are similar but not the same, and they work on distinctly different principles. In displacement, air is flooded at floor level with low-velocity air outlets. In UFAD, air is introduced at a higher velocity, which causes some induction of air at air outlets. The induction action causes still air in the room to mix with the air from the floor. UFAD has an air plenum between the floor slab and the raised floor. Floor air outlets are installed to supply air to the space. In displacement ventilation, neither a raised floor nor an air plenum is required. However, if an underfloor plenum is available, it can be used. Displacement ventilation air can be supplied from a side wall outlet located at floor level. In UFAD, the underfloor plenum is essential.

In most office spaces, UFAD is more popular than displacement ventilation. UFAD provides the flexibility to locate air outlets in office cubicles. UFAD is discussed further in a separate chapter.

48 DISPLACEMENT VENTILATION

TABLE 3–2 THE DIFFERENCES BETWEEN DISPLACEMENT VENTILATION SYSTEMS AND UFAD

No.	Displacement	UFAD
1	Air is at low velocity through the air outlet.	Air is at a higher velocity, which causes turbulence of surrounding air, leading to induction of air at outlets.
2	Underfloor plenum is not required.	Underfloor plenum is essential.
3	Air outlets are on a side wall or floor.	Air outlets are in the floor.

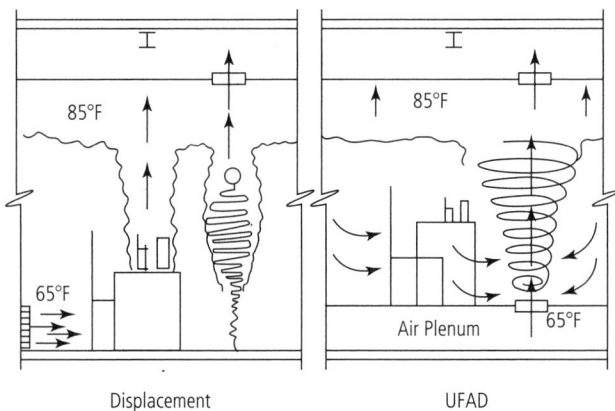

Figure 3–4 Difference between displacement and UFAD systems. *Asif Syed*

APPLICATIONS

1. Large public spaces: Cafeterias, dining halls, casino floors, train station waiting areas, airport terminal and lounges, and exhibit spaces
2. Health-care and hospitals: Patient rooms, lobbies, nurse stations, and cafeterias
3. Teaching environment: Classrooms, lecture halls, and auditoriums
4. Theaters: Auditorium and stage
5. Office buildings: Office spaces, conference rooms, and auditoriums

LARGE PUBLIC SPACES (CAFETERIAS, DINING HALLS, EXHIBIT SPACES)

Large public spaces such as building lobbies, public assembly halls, cafeterias, exhibit spaces, airport terminals, train station waiting areas, and large dining halls are best suited for displacement ventilation. Normally these places have high ceilings, large skylights, and large windows. The large amounts of glazing bring in light and views to the exterior environment, which are beneficial to the occupants. However, they also bring in a lot of heat from solar radiation. The solar radiation from glazing is a significant load. Traditional air conditioning approaches to these spaces are similar to those for office spaces, such as the mixed-air system. The efficiency of mixing systems is low in office spaces; however, it is even lower in spaces with high ceilings. One of the benefits of displacement ventilation is that the entire space does not have to be cooled. Only a fraction of the space—the occupied 6-foot zone—is cooled. A high-ceilinged space, such as a 50-foot-high hall, has about 10 feet for human occupancy, and the rest of the space is not occupied. Stratification of these unoccupied spaces with higher temperature reduces energy consumption without affecting occupant comfort. Large public spaces with displacement ventilation include the Hearst Tower designed by Norman Foster of Foster + Partners.[3] The Newseum in Washington, D.C., designed for the Freedom Forum by Polshek Partnership (now Ennead), uses displacement ventilation at the main lobby.[4]

Examples of displacement ventilation in public spaces:

Figure 3–5 Newseum lobby, Washington, D.C.   *Asif Syed*

---

[3]Nadine M. Post, "Owner Is Radiant About Lobby's Radiant Cooling," *Engineering News-Record*, Sept. 26, 2007.
[4]Newseum by Polshek Partnership Architects, 2011.

Figure 3–6 Hearst Tower lobby, New York City. *Asif Syed*

HEALTH-CARE

Hospitals and health-care facilities operate 24/7 and 365 days; therefore, they are very high energy-consuming facilities. A typical hospital uses anywhere from four to six times more energy than an office building. The energy consumed by a 1,000,000-square-foot hospital costs about $8,000,000 annually.[5] A reduction in energy of about 30 percent can be achieved with several energy-saving measures, which can save about $2,400,000 annually. When hospital systems evolved in the United States, energy was cheap, and energy consumption was not a driver influencing designers, code officials, or owners. There is no one major energy feature that will reduce energy by 30 percent; however, several smaller energy-saving measures can add up to 30 percent. Displacement ventilation is one such step that can contribute significantly to reducing energy consumption in hospitals.

Displacement ventilation can be achieved in a hospital with a central air system and local cooling units. Central air systems are similar to variable air volume (VAV) systems, but air is supplied at floor level. A central air displacement ventilation system can be

---

[5] Energy calculations for a NJ-based hospital, AKF Engineers, 2008.

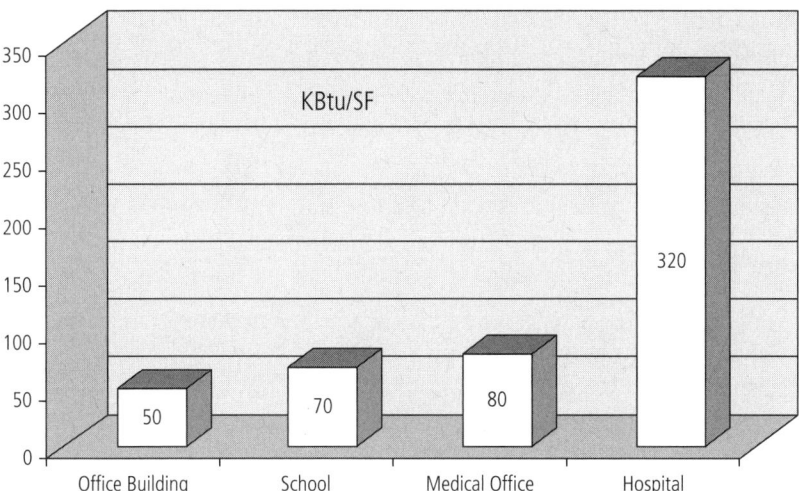

Figure 3–7 Energy use in hospitals. *Asif Syed*

used in patient rooms, lobbies, public spaces, dining halls, cafeterias, and emergency department (ED) waiting rooms. Areas with high occupant densities, which require high quantities of ventilation or outside air, are more suitable for displacement ventilation.

Displacement ventilation can be achieved with local cooling units. This practice will be discussed in detail in a separate chapter on Displacement Induction Units.

Infection control is paramount in patient rooms and emergency department waiting areas. Displacement ventilation naturally reduces the spread of contaminants and thereby helps prevent infection, as there is no horizontal or cross flow of air and airborne contaminants. Conventional ceiling air distribution works on the principle of mixing the room air thoroughly to maintain a constant temperature throughout the room. The thorough mixing of air mixes in contaminants—airborne microbes, in hospitals. In displacement ventilation, air flow is unidirectional, upward from the heat-producing object—the patient—to the air exhaust. This eliminates mixing and thereby reducing cross-contamination.

Hospital Codes

The most common standard used to design hospitals is the American Institute of Architects (AIA) guidelines for hospitals. The standard is *Guidelines for Design and Construction of Health Care Facilities* by the AIA's Academy of Architecture for Health. The guidelines do not explicitly prohibit the use of displacement ventilation. However, it is not commonly used in the United States, primarily because engineers, architects, owners, and building operators are not familiar with it. The construction industry likes to stay with what has worked in the past. However, driven by the recent increases in energy costs and greenhouse gas emissions, architects, owners, engineers, and building operators are under pressure from the CEOs of health-care organizations to address the situation. In response, engineers and architects are opening up to the idea of using displacement ventilation. Kaiser Permanente's Modesto Medical Center in Modesto, California, had a pilot program[6] to test displacement ventilation in its new facility. The hospital reports that it is able to maintain comfortable conditions with one-third less air, resulting in reduction in capital costs for ductwork and central air-handling systems. The hospital also reports that the system may improve air quality.

History of Hospital Systems

Hospitals built in the 1920s and 1930s had large windows and steam radiators. The windows helped ventilate the rooms with large quantities of outside air. The designers of these hospitals could not have foreseen the technologies of today, or even of the 1960s. During the Second World War, air conditioning companies such as Chrysler's Airtemp started to air condition hospitals. The common designs were window air conditioners and/or wall air conditioning units. Later in the 1950s and 1960s, with the introduction of large chiller systems and chilled water, the most common systems were fan coil units and induction units. Some hospitals built in the 1970s used chilled ceilings or chilled radiant panels in patient rooms. The panels also doubled as heaters in winter. The chilled radiant panels were a good energy-savings feature, but went out of fashion because of operational problems. When the chilled water temperature was not controlled, it led to condensation on the panels. Normally these panels were designed to operate with chilled water above the dew point of the rooms, but control technology in the 1970s was not sophisticated enough to predict and prevent failures, and the facilities' maintenance operators did not have the requisite education and experience in the operation and maintenance of radiant cooling systems. Some hospitals abandoned the radiant cooling systems, and replaced them with fan coil units or induction units. Some hospitals limited the radiant panels' use to heating alone and added or expanded the central air systems. Radiant cooling systems were not common in other sectors of the air conditioning industry. There are several older hospitals in business with fan coil units and induction units. Most new hospitals, from the 1990s on, are being built with central VAV systems. The use of fans makes the central VAV system the most energy-consuming system, and in some markets it is also the most expensive. The system that is the most energy consuming has become the norm. However, there are signs of change, with institutions like Kaiser Permanente venturing in a new direction by experimenting with displacement ventilation. Given rising energy costs and the demand for sustainability, the hospitals of the future will have to seriously consider new and innovative systems.

---

[6]"Model of sustainability: Kaiser's new template design comes up green," Amy Eagle, *Health Facilities Management*, April 2009. Available online at: www.hfmmagazine.com/hfmmagazine_app/jsp/articledisplay.jsp?dcrpath=HFMMAGAZINE/Article/data/04APR2009/0904HFM_FEA_CoverStory.

TEACHING ENVIRONMENT: CLASSROOM

Displacement ventilation can solve all the problems in classrooms: excessive noise, poor ventilation effectiveness, cold drafts from dumping of cold air, and poor air quality or contaminant control.

Classroom designs in K–12 schools and in colleges and universities are some of the best applications for displacement ventilation. The traditional systems do not produce an environment conducive to teaching and learning. The comfort level of occupants is poor. The energy consumed is high. The indoor air quality is poor. The acoustics are poor (noisy). All these conditions have a negative impact on the education of thousands of students who pass through year after year. Students often complain about not being able to hear the teacher over the noise. Students are not fully alert because of poor indoor air quality. Good design can have a big impact on the educations of students. A K–12 school building or a college classroom building has about fifty classrooms, each sized for thirty students. Each year about 1,500 students use these classrooms. Given the fifty-year life span of the buildings, about 75,000 students pass through this space. Imagine the impact a well-thought-out system—one that addresses all noise, air quality, and energy issues—can have on their education!

Classrooms or teaching environments traditionally were designed with air conditioning units located in the classrooms or with central systems. The most common local systems were fan coil units or packaged terminal air conditioners (PTACs). A central air system is usually a VAV system. Fan coil units and PTAC units are noisy. The traditional fan coil or PTAC unit does not comply with current noise standards. The fan coil noise is from the fan. The noise from a PTAC is even worse, as it has compressors in addition to fans. VAV system issues include noise from air outlets, poor ventilation effectiveness, dumping of cold air at low loads, and lack of contaminant control. Both VAV systems and fan coil or PTAC systems cause student discomfort problems, such as noise, poor air quality, uncomfortable hot and cold spots, and draft (dumping of cold air).

Acoustics is an important consideration in the classroom design. The HVAC system is the main source of noise and contributes significantly to the overall noise level. Most complaints in classrooms, from teachers and students alike, are that they can't hear clearly, especially in the farther rows, because of HVAC system noise. Noise is more of a disturbance in classrooms with fan coil units or PTAC units. The noise generally comes from the fan in the fan coil unit system. As the fan coil units get older, the noise from the fan increases due to the wear and tear on bearings and fan alignment. PTAC units are noisier than fan coil unit systems because the problem is compounded by the compressor in the unit. VAV systems are equally noisy, as they have to distribute large quantities of air through ceiling diffusers. In VAV systems, the acoustics can be controlled by adding duct silencers and locating the VAV boxes farther away (usually in the corridor). However, the silencers add resistance to air flow in the system, which increases the fan energy, adding to an already energy-intensive system.

According to the Acoustical Society of America (ASA), speech intelligibility in classrooms is 75 percent—that is, one in four words is not heard intelligibly. Acoustics in classrooms is a wide and elaborate subject. The booklet published by the ASA titled "Classroom Acoustics—A Resource for Creating Learning Environments with Desirable Listening Conditions" covers all aspects of acoustics in classrooms. The problem of excessive noise in classrooms from HVAC systems is almost universal; in response, standards have become more stringent. The standard commonly followed by most colleges, universities, and school boards is ANSI/ASA S12.60–2002. This chapter limits the discussion to the HVAC-generated noise. Complying with this standard is difficult and expensive with standard conventional systems such as VAV, PTAC, and fan coil units. The attenuation devices or mitigation methodologies are expensive and add costs to the project. The attenuation devices add resistance to air flow, increasing fan static, which in turn increases energy consumption and creates more noise. Mitigation methods include locating fan coils or VAV boxes far away from the classrooms, increasing the cost of ductwork and the energy used by the fans.

To conclude, the superior acoustics of displacement ventilation allow for a sustainable classroom with the least energy consumption, and the system has better indoor air quality with lower contaminant parts per million. The report prepared for the state of California titled "Advanced HVAC Systems for Improving Indoor Environmental

Figure 3–8 St. John's University, D'Angelo Building, Displacement ventilated classrooms in new construction.   *Asif Syed*

Figure 3–9 Classroom in D'Angelo Building with displacement ventilation. *Asif Syed*

Figure 3–10 St. John's University, St. John's Hall, Renovation of existing 1950s building with displacement ventilation in classrooms. *Asif Syed*

Figure 3–11 Renovated classroom in St. John's Hall with displacement ventilation. *Asif Syed*

Quality and Energy Performance of California K–12 Schools" recommends displacement ventilation as well-suited to classrooms in California. The report encourages school officials to screen the systems before deciding on an HVAC system for both new buildings and renovations.

PERFORMANCE SPACES AND THEATERS

The history of displacement ventilation in theaters goes back to 1891. Carnegie Hall in New York City was cooled with an air supply below the seats. This was before refrigeration cooling was invented. Carnegie Hall was cooled with air that flowed over slabs of ice. Large slabs of ice were dropped from the street into the sidewalk vault. A large fan blew air over the slabs, which cooled the air. The cool air was blown under the seats. Carnegie Hall was the first displacement ventilation theater. However, when mechanical refrigeration was introduced at Carnegie Hall, the displacement system was abandoned. The air supply outlets and air return outlets were reversed. The supply air outlets under the seats were changed into returns, and new diffusers in the ceiling were introduced.

Most theaters have a plenum under the auditorium. The plenum is required for the sloping of the auditorium floor to provide proper sight lines to the audience without obstructions. This plenum, which is already available, can be made suitable for airflow and used as an air supply plenum without adding any additional cost to the project. Usually, the balcony has a similar plenum. Some of the considerations of this design are:

- The plenum is 10 to 12 inches high at the first row. The increase in the plenum height due to standard sight line sloping is sufficient for air distribution. Additional sloping for air distribution is not required.
- The concrete plenum surfaces are sealed off to prevent dust generation.
- All joints at the floor–wall corners and other joints must be sealed airtight, to avoid airflow losses.
- There are no obstructions to airflow in the plenum. The structure that supports the plenum has to be open to allow for airflow. Careful coordination among the structural engineer, architect, and mechanical engineer is required.
- The depth of the plenum at its widest has to be deep enough for a low air velocity. Low air velocity is essential for proper functioning of the displacement system.

Air outlets: The type of air outlet used in the displacement ventilation system is the 8-inch round floor-mounted unit. This air outlet is commonly referred to as a salad spinner, because it resembles just that. When installed, the air outlet is flush with the floor or the carpet. This type of air outlet is similar in appearance to diffusers used in the office underfloor air distribution (UFAD) system, but it is distinctly different in performance. The UFAD diffuser induces air from the room, but with turbulence, which means higher velocity and higher noise. The displacement diffuser has no induction effect, and the velocity of air flow is slow, and therefore almost silent. You may have noticed that for the displacement system I have used the term "air outlet," and for UFAD I have used the term "diffuser." The difference is that an air outlet just allows the air to spill out or overflow, and a diffuser ejects or forces the air outward. The quantity of air outlets varies based on the heat load of the theater and the acoustic criteria to be achieved. Generally, one diffuser for every two seats meets the requirements for air flow as well as heat load for most theater applications.

Return air: The ideal location for return air is in the ceiling; sometimes openings in the ceilings for spotlights can be used. In theaters where there are no ceilings, the return ducts could stub out at the highest point. It is important to provide return air at several spots in order to minimize the air velocity noise from the return air ducts, especially in theaters without ceilings. It is preferable that air-handling units be located as far as possible from the auditorium and ceiling. The duct lengths from the air handler to the supply plenum below the auditorium and the return air duct in the ceiling help in attenuating the noise generated by the air-handling unit.

Lighting: Lighting in most auditoriums is located in the ceiling, along the side walls, and at the edge of the balcony. The intensity of lighting is very high. Sizing the air-handling units and the ducts and air distribution system to handle the heat generated by the most intense lighting is generally not necessary. Most of the high-density loads occur only for brief periods. It is also important to note that all lights do not come on at the same time, and there is reasonable diversity. Displacement ventilation works perfectly for this purpose, because heat loads from lights are in the return air path, after the air has cooled the auditorium's occupied zone. Most of the lights are directed onto the stage, and the radiant portion of the light heat loads to the auditorium is relatively small. In the conventional air distribution methods, the heat from the lights heats up the air before it comes down on the audience.

Acoustics: It is needless to point to the importance of acoustics in theaters and performance spaces. In a conventional system, the air has to be delivered to the auditorium floor from the high ceiling. In order to diffuse the air evenly to the entire space, some type of ceiling diffuser is used. Most ceiling diffusers discharge air at relatively high velocity, which is the main source of noise. In a displacement ventilation system, the velocity of the air is very low, reducing the noise levels to the bare minimum. The acoustic standard for performance spaces is very high, in the range of NC 15–20 or dBA 25–30. NC 15 is the threshold of hearing. Recently the standards have become more stringent and call for lower NC and dBA levels; older books have NC 20–25 as standard for theaters and performance spaces. With such high demand for lower noise levels, displacement ventilation is one sure way of reducing noise from the air conditioning system.

Stage: Air conditioning systems for the stage are generally not combined with the auditorium system, especially in public performance theaters. However, in small college and university teaching and light performance theaters, the stage and auditorium systems can be combined. The reasons for not combining them are: The heat loads to cool are generally different in the stage and the auditorium. The stage usually has different operational hours for practice and rehearsals, when the entire theater does not need to be cooled. Displacement ventilation systems are generally not suitable for performance stages because there will be movable scenery in the way of air flow. The location of scenery changes from performance to performance, thus making it difficult to establish fixed locations for air outlets. Therefore, the choice for the stage is generally an overhead air system. When the stage system is combined with the auditorium system, as in the case of college and university teaching theaters, it is important to note that a displacement system for the auditorium cannot be combined with an overhead system for the stage. The overhead system supply air temperature is 55°F, whereas the displacement system is 65°F. One unit cannot supply two different temperatures.

Example: Samuel J. Friedman Theater for the Manhattan Theater Club (formerly known as the Biltmore Theater), New York, uses displacement ventilation. The theater

Figure 3–12 Samuel J. Friedman Theater for the Manhattan Theater Club—Exterior, uses displacement ventilation.   *Asif Syed*

was originally built in the 1920s, and the original design did not include an air conditioning system. A conventional air conditioning system was later added with an exposed duct. The theater was closed down due to a fire, and in late 1990s it went through a renovation, during which a displacement system was added. The space below the auditorium and balcony seats was used as a plenum. Additions of displacement ventilation brought back the original historic and landmarked character to the building, without any changes to the finishes. The air conditioning system was not visible in the finishes.

OFFICE SPACES

Displacement ventilation has some limitations in an open-plan office layout. UFAD, which is similar, is better suited to the office environment. Displacement ventilation requires air to flow across the floor, which is prevented by cubicle office partitions, making the system impractical for open office layouts. However, in office spaces where air can flow across the floor, displacement ventilation can be used. Some new office layout designs are abandoning the cubicle structure with partitions going down to the floor. A more open environment with just tables and chairs where interactive and collaborative team members sitting opposite, across, and side by side from each other is observed and preferred by some companies. In this environment, there is no obstruction to air flow across the office space at the floor level. Displacement ventilation can be adopted in these designs.

For most office spaces, UFAD is more popular than displacement ventilation. UFAD provides the flexibility to locate air outlets in cubicles. UFAD is discussed in a separate chapter.

Corporate headquarters auditorium: Most corporate headquarters buildings have an auditorium with a small stage. Displacement ventilation works well for these auditoriums. If available, an underfloor plenum can be used. Side wall air outlets work equally well. Because these auditoriums are used infrequently, it is always advisable to install a separate dedicated system, which can be shut off.

CHAPTER 4

Chilled Beams

CHILLED BEAMS HAVE CAUGHT THE ATTENTION AND CURIOSITY of many in the building design and construction industry. Their popularity is growing in the United States. Several new buildings and retrofits of old buildings are using this technology. Several manufacturers have started manufacturing chilled beams in the United States. Only a few years back, they had to be imported from Germany or Sweden. Carrier Corporation, a leading U.S. manufacturer, has started offering chilled beams in their product line.[1] Several new projects have started using chilled beams. Buildings at institutions of higher education were the first to use them in their labs and offices; lately, commercial office buildings have started using them as well. Many sustainable design and LEED-certified projects have incorporated chilled beams as a means to improve energy efficiency and comfort, especially since the LEED certification process has increased its emphasis on energy efficiency, and mandates that energy efficiency points are necessary for certification. Despite their name, "chilled beams" have nothing to do with the beams or structural elements in a building. Chilled beams got their name from their slender, long shape, which was similar to that of structural concrete beams. When installed in buildings without ceilings, they were mounted to the underside of the slabs, where they looked like structural beams. The modern chilled beams do not look like beams and are not limited to use in spaces without ceilings, but the name has remained. There are several types of chilled beams available, and they can be used in various types of architectural configurations, with ceilings or without ceilings.

---

[1] Carrier Corporation, press release: "Carrier Corp. Introduces New Chilled Beam Cooling Products," January 22, 2008.

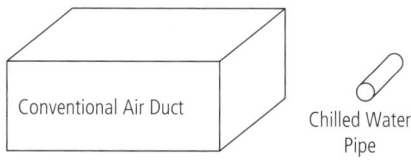

Figure 4–1 Air- and water-based cooling *Asif Syed*

30 tons of cooling (sensible)
16,000 CFM of air = 12 square feet of duct
72 gallons of water = 3 inch diameter pipe
Ratio 266:1

Chilled beams are associated with sustainability and energy savings. They have been popular in the European building market since the 1970s; the technology was primarily developed to meet the need for air conditioning in the building retrofit market. Older buildings in Europe were not designed with air conditioning. Given the increased demand for air conditioning, they worked well in existing buildings that were retrofitted. But they have one other thing going for them: energy savings. Compared to conventional all-air systems such as VAV, they consume less energy to achieve the same level of comfort. This made the technology very attractive, and chilled beams found their way into the new building market.

PRINCIPLE OF OPERATION AND TECHNOLOGY

The technology of chilled beams is the same as that of the all-too-familiar induction units. Chilled beams are sometimes referred to as ceiling induction units. Both chilled beams and induction units work on the same principle, but the performance of present-day chilled beams and induction units is far superior. They are quiet and operate at much lower air pressure. The principle of induction is to induce air flow, while a jet of air comes out from high pressure to low pressure. The history of this technology can be found in the United States. All building professionals are familiar with induction units in buildings in the United States. Induction units were standard perimeter heating and cooling elements in buildings in the 1960s and 1970s. Most of the high-rise office buildings in New York, Chicago, and Boston had them. The World Trade Center in New York City had 24,000 units in both towers.[2] The units went out of fashion as glass performance improved, taking a major leap with shading coefficients, light transmittance, thermal breaks, and insulating values. With such high-performance glass, air distributed from ceilings was sufficient to maintain comfort. Heating from under the windows, which was the standard practice, was not required anymore. This development was welcomed by the real estate industry, as it saved 15 to 18 inches of floor space along the perimeter of the building—precious rentable floor space. At the same time, while the use of induction units was decreasing in the United States, it was increasing in Europe. Several companies in Europe researched the induction principle of air distribution and developed the active chilled beam, the most popular among the various types of chilled beams. One major development from the 1960s to the 1990s was that the pressure at which chilled beams can work dropped from 6 inches to 1.5 inches. This reduced pressure of operation and development of design led to a quiet operation. The high-pitched induction unit noise of older buildings is absent when chilled beam and other modern induction units are used. Chilled beams have a variety of applications. The most popular application is in office buildings and laboratories.

BENEFITS OF CHILLED BEAMS

The benefits of chilled beams are:

1. Low energy consumption
2. Better comfort and lower noise
3. Space savings
4. A flexible system for high churn rate
5. Low maintenance, no moving parts

ENERGY SAVINGS

The energy savings from chilled beams are realized from the way heat is carried from the space (source) to the outdoors (sink). In a conventional system, air is the carrying medium. Air is cooled in a cooling apparatus, which is usually an air-handling unit, and circulated with a fan. The cold air carries its cooling capacity to the space and is circulated back to the space. In a chilled beam system, the inefficiencies are overcome as follows:

---

[2] *Engineering News-Record*, 1967.

1. Fan inefficiency: Fan systems are large pieces of equipment that are inefficient and take up space. The standard fan efficiency is 65 percent; the remaining 35 percent is lost as heat gain. This is heat added to a system that has been cooled for air conditioning. The air that is passed through a fan has passed through a cooling coil, if the coil is upstream, or will pass through the cooling coil, if it is downstream. Either way, the fan adds heat to the air, which has to be cooled, making the system inefficient. The energy required to move air is a significant portion of the overall building energy. In chilled beam technology, the fan system is reduced in size by about 60 to 70 percent, which significantly reduces the fan energy.

2. Air inefficiency: Air is not an efficient medium to carry heat. A good example of this is the fact that one can stand in 65°F air, but not in a pool of 65°F water. Water carries away the heat much faster than air. Since air is an inefficient medium to transfer heat, large quantities or volumes are required, which means that large fans with large motors are required. In chilled beam technology, water is the heat-carrying medium. Water is an excellent heat-carrying medium—for example, on a hot summer day, dipping in the pool gives more comfort than fanning with a large fan. The high heat-carrying capacity of water requires only low volumetric flows through the building. Water pumps are far more efficient than air fans. The efficiency of water pumps is 85 percent (compared to 65 percent for air fans). The pumping energy for a water system is less than for an air system.

3. Chiller energy savings: The chilled water temperature used in chilled beams is higher than for the conventional systems. Conventional systems use 45°F chilled water, whereas chilled beams use 55°F chilled water. Not having to lower the temperature to 45°F makes the chiller use less energy. In addition, when an economizer is used—that is, when the outdoor temperature is used to produce chilled water—there are more hours of free (without chiller operation) chilled water. The size of the chiller plant for chilled beam systems is the same as for a conventional system; however, the operation of the plant is more efficient. The chillers run for fewer hours, and when they run, they use less electricity.

4. Boiler energy savings: As with the chiller operation, the hot water operation of chilled beams uses lower hot water temperature. Conventional systems use 160°F water, whereas chilled beam systems use 110°F water. Not having to heat the water to 160°F makes the boiler operation more efficient.

5. Coupling with geothermal: The energy savings of chilled beams can be enhanced further when they are coupled with geothermal systems. The higher chilled water temperature of 55°F lends itself to the use of free renewable

energy available in the Earth, where year-round the temperature is 55°F. Just by pumping water through the Earth in a geothermal well, the running of the compressor can be eliminated.

COMFORT AND NOISE

The comfort and noise levels are significantly improved over those of conventional systems as follows:

1. The indoor air quality is guaranteed at all times and at all load conditions. The primary air that drives induction air has to be maintained all the time, irrespective of internal load. In a conventional system, when the load is reduced, the air flow is reduced, risking the reduction of ventilation air.

2. Constant air circulation with varying temperatures reduces drafts experienced in most conventional systems when the air flow increases or drops, especially at the perimeter of the building, due to changes in envelope and solar loads.

3. Low noise levels: Chilled beams do not have fans, which are the main sources of noise in the space. The chilled beams, when set at low air pressures, have a very low noise level. Chilled beams also eliminate the noise from the air outlets of conventional systems. Conventional system diffusers or air outlets have to operate between high and low air flow, based on the internal loads and external heat loads. The external loads vary from season to season, with high loads in summer to no load in winter. The daily changes include sun path. Afternoon west façades have higher loads at 3 to 4 PM and other times lower loads. High air flows are required at higher loads and lower air flows at lower loads. At high air flows the outlets are noisy and at low air flows the outlets dump air leading to cold drafts. Designers generally choose a balance, which does not satisfy any conditions. Chilled beams do not have this problem, as there is constant flow all the time. Constant air flow at all times permits the chilled beam to be as designed for the desired noise criteria. With chilled beam induction units, the extremely stringent noise level requirements for classroom acoustics per the ANSI/ASA S12.60–2002 Classroom Acoustics Standard can be met.

SPACE SAVINGS

Chilled beams occupy less space in buildings, compared to conventional systems. This can lead to reduction in the cost of buildings. There are two ways in which chilled beams reduce space requirements: (1) less floor space is required for mechanical rooms and shafts, and (2) less volumetric space is required in the ceilings. Both of these can reduce the construction costs. The lower volumetric space in the ceilings can reduce the slab-to-slab height, reducing the building envelope. Chilled beams are the preferred solution for buildings with lower slab-to-slab heights. The present-day ceiling requirements for most office buildings are 9 to 10 feet, while in the past this was only 8 feet. There are several existing buildings built from the 1960s to the 1980s, where the ceilings are only 8 feet and slab-to-slab heights are 12 feet. When these buildings are repositioned for the current market, 10-foot ceilings cannot be achieved with all air or conventional air distribution systems. Chilled beams have become the preferred choice in this situation at several buildings.

1. Smaller fan rooms and shafts: Chilled beams use water as the medium to transport heat, whereas conventional systems use air. The air duct system, which is limited to ventilation air, is 60 to 70 percent smaller than the conventional, all-air system. This translates to smaller air shafts and fan rooms.

2. Ceilings space: Reduction in duct sizes leads to smaller ducts in the ceilings. For the same floor area, the ceiling space can be smaller by 6 to 9 inches. This reduction can result in smaller slab-to-slab heights. This effect is significant in high-rise buildings; savings of 6 to 9 inches in a 50-floor building can result in the addition of two to three floors, or an increase in the floor area of from 4 to 6 percent. In a low-rise, five-floor building, this reduction can result in reduction in the height of the building by 30 to 45 inches, or 4 to 6 percent, resulting in savings.

FLEXIBLE SYSTEM FOR HIGH CHURN

Churn is defined as the percentage of fit-out floor area that is no longer useful because a tenant or user has left the building, or because the tenant's or user's needs have changed, and the space has to be refitted out. This is a very important matrix in a building's ongoing costs. Churn cost varies from building to building, but for some buildings or companies it can be very high. It is not unusual for the same tenant to change a space three to four times in 10 years, for varying business needs and styles. In this situation, usually the space is gutted, replacing ceilings, carpets, HVAC, electrical, and fire protection. Chilled beams offer a flexible layout with plug-and-play connections; the units can be easily relocated with just piping connections. There is no electrical work associated with relocating chilled beams, as chilled beams do not have an electrical power connection.

LOW MAINTENANCE

Chilled beams do not have any moving parts and therefore break down less. The only maintenance required is the cleaning of the nozzles. This is accomplished with a vacuum cleaner and brush. Additionally, cooling coils and wire-mesh filters require vacuum cleaning to remove dust. All other surfaces of the chilled beam are wiped with a moist cloth for removal of dust. Low maintenance leads to savings in operational costs. Lack of condensation makes the cooling coil operate dry; there are no wet surfaces.

TYPES OF CHILLED BEAMS

There are two main types of chilled beams: passive and active. As the name suggests, passive chilled beams work by the natural convection of air, and active chilled beams use the induction principle to circulate air. Passive chilled beams can also be referred to as static chilled beams. Active chilled beams are also referred to as ventilated chilled beams or ventilated induction units, since they are connected to a ventilation air system. A third type of chilled beam is a subcategory of the two types and incorporates other services, such as lighting and sprinklers. Chilled beams that incorporate other services in addition to cooling are called multiservice chilled beams.

PASSIVE CHILLED BEAMS

Passive chilled beams are the simplest form of chilled beams. Their physical appearance is similar to that of a baseboard radiator, fins with water tubes. The fins are more densely packed and look similar to a water-cooling or water-heating coil. Chilled beams are made up of copper coils or tubes with aluminum fins. They are cooling coils that are suspended in the ceilings or under the slab, where there are no ceilings. Chilled beams are connected with a chilled water system. In almost all passive chilled beam systems, air still needs to be supplied to the space through a duct system. However, this duct system is much smaller than that of the conventional all-air system. A duct system is not required in applications where ventilation can be achieved with operable windows and when the cooling load is small. Condensation is a concern with chilled beams, especially with operable windows; proper design of controls has to be incorporated to prevent condensation. When chilled beams are used in buildings where operable windows are available, it is recommended that an anticondensation control scheme be installed. The cooling capacity of passive chilled beams is small, and the applications where it can be used without an air supply are limited. Passive chilled beams cool the space by means of the natural convection of air. The warm air from the space rises up and accumulates in the area above the ceiling; the transfer is facilitated by a grille or openings in light fixtures. The warm air drops down through the cooling coil, then back into the space. The coil operates dry and does not have any condensation, because the chilled water supplied is above the dew point temperature of the space. It is recommended that the humidity and dew

Figure 4–2 Passive chilled beam. *Asif Syed*

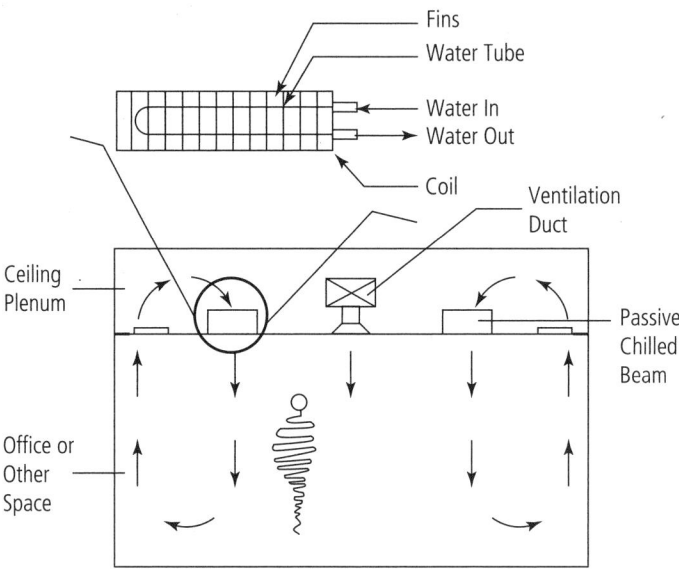

point of the space be continuously monitored via a control system, to maintain the chilled water temperature above the dew point. This guarantees dry operation of the beam and is the anticondensation control. Some of the benefits of passive chilled beams are:

1. They do not require any air or electrical connections.
2. They can be installed in spaces with high equipment loads (sensible heat).
3. They work well at the perimeter of the building to offset solar heat gain through glazing. Warm air from the windows rises up and is cooled by the chilled beam, by natural convection. A conventional ventilation system will require air circulation from a central fan, which is located at a distance and requires fan motor energy to circulate air.
4. They can be retrofitted in existing buildings to offset increases in cooling loads.

ACTIVE CHILLED BEAMS

Active chilled beams are the most common type used in buildings. The main advantage of active chilled beams is that they put the ventilation system to use to increase the cooling capacity. They have a higher cooling capacity than passive chilled beams. Active chilled beams have a duct connection similar to that of traditional ceiling diffusers. In an active chilled beam system, the air that comes out into the space is about three times the air that goes in from the duct connection. This happens as a result of the induction effect. The air that enters the chilled beam (one-third) induces two-thirds more air from the space. The induction effect is produced by the nozzles of jets inside the units. This is the reason that these units are also called ceiling induction units. As the primary air comes out of the nozzle, it induces more air flow from the area surrounding the nozzle. The air induced into the chilled beam is about 60 to 70 percent of the primary air. The induced air is also called the secondary air. The total air that comes out of the chilled beam is the sum of the primary air and the secondary air. The induced, or secondary, air is made to pass through a cooling coil that cools and provides additional cooling capacity, compared to passive beams. The chilled beams are dry; there is no condensation in the coil. This makes them clean and reduces maintenance. The induction principle is well suited and adaptable for room air distribution for the following reasons:

1. The unit produces air flow in the space without the use of a fan. The fan inefficiencies are eliminated.
2. The additional cost of the electrical power connections, which includes high field labor, is eliminated.
3. The induction principle uses the primary air for ventilation, partial cooling, moisture removal, and the secondary air or induced air for cooling. As discussed earlier, the air required for ventilation is far less than the total air supplied in a conventional system. This eliminates the circulation of large volumes of air in the building. The benefits of reducing the size of large central systems are:
   a. The reduced amount of air flow saves in fan energy.
   b. Active chilled beams reduce the amount of ductwork, which reduces cost.
   c. The reduction in ductwork allows for higher ceilings.
   d. With smaller ductwork, slab-to-slab height can be reduced, resulting in savings in building façade construction. In high-rise buildings, additional floors can be added, providing more floor area with the same height.
   e. Where zoning ordinances limit the height of buildings, such as in Washington, D.C., chilled beams permit an increase in ceiling heights.
   f. In high-rise buildings, the lower ceiling height requirement can add up and may provide an additional floor, providing valuable floor space for the same façade area.
4. Dry, or noncondensing, operation reduces maintenance or regular cleaning of drain pans in condensing units such as fan coil units.

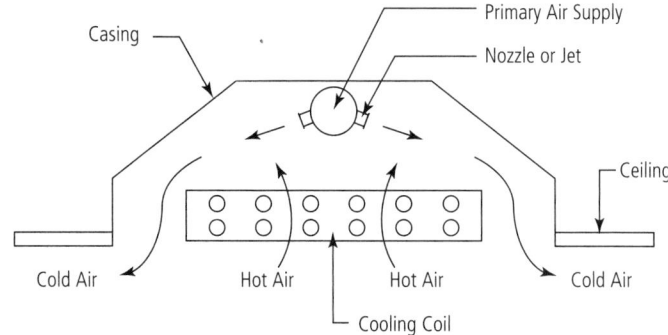

Figure 4–3 Induction effect of nozzle. *Asif Syed*

Figure 4–4 Active chilled beam. *TROX Technik*

MULTISERVICE CHILLED BEAMS

Multiservice chilled beam installations incorporate other services such as light fixtures and sprinklers into the chilled beam. Other services that can be incorporated into the chilled beam are: public address systems, smoke detectors, and CCTV cameras. Multiservice chilled beams allow for modular installation of all services to a limited area, and the rest of the ceilings are freed up. Multiservice beams can also eliminate the ceiling construction, especially in concrete construction. All services to the chilled beams can be channeled through a limited ceiling at the core, eliminating ceilings in the majority of the areas. Multiservice chilled beam installations are common in Europe. In the United States, the installations are limited in number, but growing fast. Most unions that install the units are trade-based and control the trade equipment from receiving on-site to installation and startup. In the installation of a new multiservice piece of equipment that involves electrical and mechanical trades, the unions will be unfamiliar with the unit, and the handling of the equipment and installation must be discussed with the unions and work divided up well in advance of the contract. Benefits of multiservice beams are:

1. Ceilings can be eliminated, reducing construction costs.
2. With concrete construction, exposed slab can increase the thermal inertia of the building, which reduces the cooling load demand for instantaneous loads.

3. Floor-to-floor heights can be reduced.
4. Multiservice beams are good applications in older buildings with concrete structural slabs and low floor-to-floor heights.
5. The need for complicated on-site multitrade coordination is reduced.
6. Factory assembly of multitrades reduces on-site construction time.
7. Easy plug and disconnect allows for churn of the space.
8. The technology provides opportunity for integrated design.

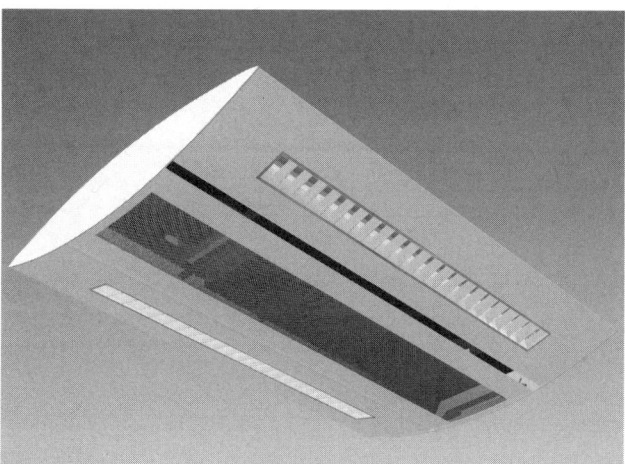

Figure 4–5 Multiservice chilled beam. *TROX Technik*

Figure 4–6 Chilled beam installation (in ceilings). *Jeff Wolfram Photography, © JeffWolfram.com*

Figure 4–7 Multiservice chilled beam (without ceilings). *TROX Technik*

CHILLED BEAM APPLICATIONS

Chilled beams are well suited for a wide variety of applications, but fundamentally they are cooling devices. Chilled beams are best suited for applications with a lot of equipment heat (also known as sensible heat). To use chilled beams the building has to have a chilled water system. Chilled beams work as local cooling devices using chilled water. Chilled beams also require a central ducted ventilation system, as the driving force for air flow around the chilled beam is the primary air. The common applications are:

1. Commercial office applications
2. Hospital patient rooms
3. Laboratories (both dry and wet labs)

Chilled beams are less suited for applications where moisture is generated (also known as latent loads). Additionally, people-intensive spaces are not suitable for chilled beams, such as:

1. Large conference rooms, auditoriums, theaters
2. Places of assembly, train stations, ball rooms, dining halls

To-do list for chilled beam design:

1. Reducing the external heat load of the building can reduce the number of chilled beams required. The external heat load can be reduced with:
   a. High-performance glazing, with double-glazed glass selected specific to the climate.
   b. Low thermal heat bridges with insulated spacers in glazing assemblies. Generally the overall insulation value of the glass plus frame is twice as that of glass. The heat transfer through the frame has a major impact in the external heat loads.
   c. Tight envelope can reduce infiltration, which is a source of heat and moisture. Infiltration of moisture is not desirable in chilled beam installations due to condensation.
   d. External solar shades, where possible, and internal shades are desirable to reduce solar gain.
2. Humidity control strategy includes:
   a. Continuous outdoor and indoor humidity monitoring must be done in all chilled beam installations. Monitoring of internal humidity and temperature is used to calculate the chilled water temperature. This is done by the building automation system.
   b. Building automation controls reset the water temperature to avoid condensation. Almost all chilled beam installations require two chilled water circuits: one for the central air system for the primary air, and the other for chilled beams. This chilled water circuit for chilled beams is generally at 55°F, where as the central or primary air circuit is at 45°F. A plate and frame heat exchanger can be used to accomplish this. Sometimes the water leaving the primary air system circuit is used for the chilled beam circuit. If this is done then it has to be verified that the water balance is at all load conditions.
3. Layout of chilled beams:
   a. Develop a modular ceiling layout of beams coordinated with modular office layout.
   b. Additional beams are required at the perimeter to offset solar gain. Reduction in solar load, as noted in (1), helps reduce the quantity of beams.
   c. When passive chilled beams are used, they have to avoid installation directly over occupants to avoid downdrafts.
4. Active vs. passive beams:
   a. Carefully analyze the load capacities and the right match of active or passive beams.

b. Most applications for normal commercial-type spaces require active beams.

c. Low-heat load spaces, or spaces that require supplemental cooling, are suitable for passive beams. Applications with high loads such as labs are suitable for active chilled beams.

5. Ceilings:

   a. Ceilings are generally used in most chilled beam applications, but they are not essential. Both active and passive chilled beams can be installed in spaces without ceilings. Multiservice chilled beams can be installed in new spaces where ceilings can be limited to a central core and not extend up to the building perimeter. Evaluate the need of ceilings, type of chilled beams, ceiling heights required, and so forth.

   b. If ceilings can be avoided, multiservice chilled beams can be used.

6. Economics:

   a. Perform a life-cycle cost analysis of the installation for 30 years.

   b. Include the churn cost downside of conventional systems and the upside of chilled beams. Only a complete and accurate analysis will determine if there is an additional cost. If there is an additional first cost, then life-cycle cost will prove the payback period and also the savings over the entire period.

7. Perimeter heating system:

   a. Identify the type of heating at the perimeter in regions where heating is required.

   b. Four-pipe chilled beams are available for heating and cooling.

   c. Electrical heat at chilled beams is normally provided in the air stream of primary air.

   d. Electrical heating chilled beams are in development now to address the market where electric heat is popular. In several regions, the cost of electricity may be low, gas infrastructure may not be available, and 8 to 10 hours' use per day does not pay in boiler investment. In these instances chilled beams with electric heat can be used.

8. Humid climates:

   a. Active chilled beams can be used in all climates, including humid climates. For all chilled beam installations, humidity control is given special attention.

   b. The ventilation air is generally sufficient to remove humidity generated by occupants.

c. Humidity due to infiltration has to be controlled with a tight envelope.

d. Building system operations are important. Morning startup must include dehumidifying the building prior to sensible cooling. Nighttime operation of exhaust fans must be avoided to prevent humidity infiltration. In high-rise buildings, the stack effect can infiltrate air into the building. This infiltration can bring in humidity from outdoors. Special measure must be taken to reduce the stack flow in chilled beam high-rise buildings.

COMMERCIAL OFFICES

Chilled beams are not new to commercial office buildings. They have been used in Europe since the 1980s. There are several high-profile projects that have successfully used chilled beams. The primary driver behind the use of chilled beams is their energy savings, sustainability, and flexibility. They can be used in new buildings and also be retrofitted in old buildings.

Installations in the United States are limited in number, but growing at a very fast rate. One of the largest installations in the United States is the Constitution Center in Washington, D.C. This 1.4 million square feet of commercial office space uses chilled beams.[3]

Figure 4–8 Constitution Center, Washington, D.C., chilled beam installation. *TROX Technik*

---

[3] Constitution Center website, http://constitutioncenter.dc.com. Constitution Center, 400 7th Street SW, Washington, D.C. 20024.

There are several different ways chilled beams can be used in commercial office buildings:

1. Active chilled beams with overhead distribution
2. Passive chilled beams with UFAD for ventilation
3. Passive chilled beams at perimeter to offset solar and envelope heat gain with UFAD or with overhead air distribution

The most common type of chilled beam installation used in office spaces is active chilled beams. The active chilled beams incorporate ventilation air connections. Active chilled beams can meet and exceed the ventilation requirements of most office building installations. Ventilation air, or the primary air required by chilled beams, can also meet the excess ventilation required for LEED certified projects. Chilled beams come in several sizes and cooling capacities. The most common are 6 feet and 4 feet long. The longer chilled beam has much higher capacity, less piping, and fewer duct connections and controls. The quantity of chilled beams required in the office space is based on the occupant density and the electrical power use (heat generated) in the space.

The standards for most commercial office buildings specify about 5 watts per square foot of cooling capacity. Most leases require that the landlord provide a cooling capacity of 5 watts per square foot for heat generated from office equipment and lights. This 5 watts per square foot criteria has been used from the 1990s and is still in use today. However, the energy efficiency standards for office equipment and lights have significantly improved. The current code for lights does not permit more than 1.0 watts per square foot (compared to 1.5 watts per square foot in the 1990s). The use of LCD screens has also reduced office equipment power use. All electrical power used in the space is converted into heat. Except for a few exceptions of special buildings, generally in most office buildings the total power use is about 3 to 4 watts per square foot. The 3 to 4 watts per square foot is based on:

1. Standard density of occupancy of one person for every 100 to 125 square feet
2. Standard office equipment, one standard computer per person
3. Average computer use or draw of 200 watts (these computers may be rated much higher at 300 to 400 watts) per person use
4. Average LCD use or draw of 50 watts
5. 1.0 watt per square foot of lighting density in the ceiling

Table 4–1 shows the methodology to calculate the office power consumption of a standard office space, for standard office building capacities of 3 to 4 watts per square foot. High equipment power loads can be realized in offices with higher equipment loads, such as trading floors, where density can be as high as 7 watts per square foot. Table 4–1 does not include moisture (latent) load calculations.

TABLE 4–1 OFFICE-SENSIBLE (NONMOISTURE) SPACE HEAT LOADS

Density	100 Sq. Ft./Person	150 Sq. Ft./Person
Occupant load	50 watts	50 watts
Lighting load	100 watts	150 watts
Computer load	200 watts	200 watts
Monitor load	50 watts	50 watts
Total	400 watts	450 watts
Watts/sq. ft.	4.0 watts/sq. ft.	3.0 watts/sq. ft.

TABLE 4–2 STANDARD OFFICE AT 4 WATTS PER SQ. FT. ACTIVE CHILLED BEAM COVERAGE

Beam Length	Floor Area in Sq. Ft.	Maximum Occupants
4 feet	250–450	2–5
6 feet	450–750	3–8

TABLE 4–3 HIGH LOAD DENSITY OFFICE AT 7 WATTS PER SQ. FT. ACTIVE CHILLED BEAM COVERAGE

Beam Length	Floor Area in Sq. Ft.	Maximum Occupants
4 feet	150–250	2–5
6 feet	250–450	3–8

The moisture generation from people is removed from the space through ventilation air. Chilled beams provide cooling for only equipment (sensible) heat. Separate calculations for moisture removal are required. Moisture removal calculations are not shown here, as the essence of this chapter is to demonstrate the use of chilled beams with office loads. Chilled beam capacities can vary based on the manufacturer, chilled water temperature, configuration, primary air quantity and temperature, nozzle design, and so forth. Tables 4–2 and 4–3 can be used as a rule of thumb to estimate the quantity of chilled beams required. The coverage shown in Tables 4–2 and 4–3 are for office interior spaces. The external perimeter spaces require more or higher-capacity chilled beams due to envelope loads. The envelope loads are generally limited to a 15-foot perimeter zone. Chilled beam coverage for both standard office space and higher-load office space is shown in Tables 4–2 and 4–3, respectively.

CHILLED BEAM USE WITH UNDERFLOOR AIR DISTRIBUTION (UFAD) APPLICATIONS

Chilled beams can be used with underfloor air applications. In UFAD and other systems the loads at the perimeter are variable due to solar heat gain, which varies seasonally and also daily with the path of the sun. The peak solar load occurs for only a few hours in the entire year, usually in the summer months. Chilled beams along the perimeter can absorb the solar heat gain, eliminating that load from the central UFAD system. Therefore, in UFAD applications, the varying perimeter space load from the envelope and solar heat gain can be offset with chilled beams. This reduces the duct and fan capacities of the UFAD system, reducing the size of the duct shafts, fan rooms, and plenum depth. This optimizes the UFAD system, which includes the fans, ducts, and air plenum to the size that runs at peak capacity for the majority of the time. Active chilled beams are available with one-way throws or air distribution that can be used along the glazing. Passive chilled beams can also be used at the perimeter zone. Passive chilled beams do not require any duct connections. This eliminates the supply duct system in the ceiling, making passive chilled beams more cost effective.

Passive chilled beams along the perimeter are the preferred choice over active chilled beams. Passive chilled beams do not require any duct connections, making the installation simple and easy. The building envelope has to be integrated with chilled beams. Coordination between architect, façade consultant, and engineer is required to match the cooling capacity of chilled beams to that of heat gain from the envelope. The reduction of the building envelope load through high-performance

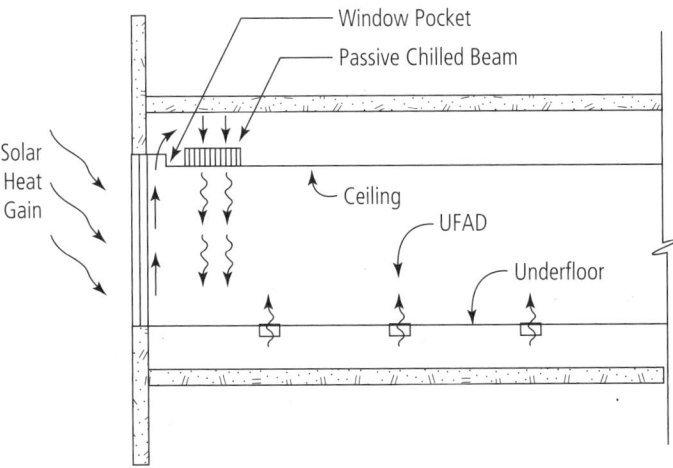

Figure 4–9 Passive chilled beam in UFAD application. *Asif Syed*

glazing and envelope can significantly reduce the number of chilled beams required. This application reduces the energy use, and improves overall thermal comfort in the building.

HOSPITAL AND PATIENT ROOMS

Chilled beams work on the principle of induction or induction technology. Induction technology is not new in hospitals; it has been used in several hospitals built from the 1950s to 1970s in the United States. Several hospitals in New York City have induction units. Induction units are normally located under the window sill in the patient rooms and in the administration areas. Induction units do not have a fan and circulate the room air through induction created from the primary or ventilation air forced through the nozzles. The primary reason for the use of induction units was to avoid recirculation of air in the hospital patient rooms. Recirculation creates cross-contamination between different spaces in the hospital. A conventional 100 percent outside air system was too elaborate (three times the size of induction) and energy intensive, even with the low cost of energy in the 1960s, so induction units provided an excellent alternative. Induction units offered the opportunity to reduce the size of the ducts, reduce energy use, and provide zone control. The reasons for using chilled beams are still the same, but the performance of chilled beams has significantly improved from the 1970s' induction units technology. The advances include operation at lower pressure and nozzle designs that are quieter.

A common problem encountered in hospitals during the design and construction is the lack of ceiling space to install ducts. This problem is more severe in central air systems such as VAV. All the ducts have to be channeled through the corridor ceilings. The VAV boxes are also located in the corridor to avoid maintenance from patient rooms. A large volume of space is required just over the corridor ceiling. To get this volume there are two options: lower the ceiling height or increase the slab to slab. The lower corridor ceiling is architecturally not desirable. Increasing the slab-to-slab height just for corridor ceiling increases the cost of construction. Chilled beam systems reduce the size of the ducts in the corridor to one-third of the central air system.

The benefits of chilled beams are:

1. They can be installed in the ceilings, freeing up the floor space.
2. They are quieter than induction units.
3. They offer better zone control.
4. They save energy.
5. They reduce the size of the central fan and duct systems.

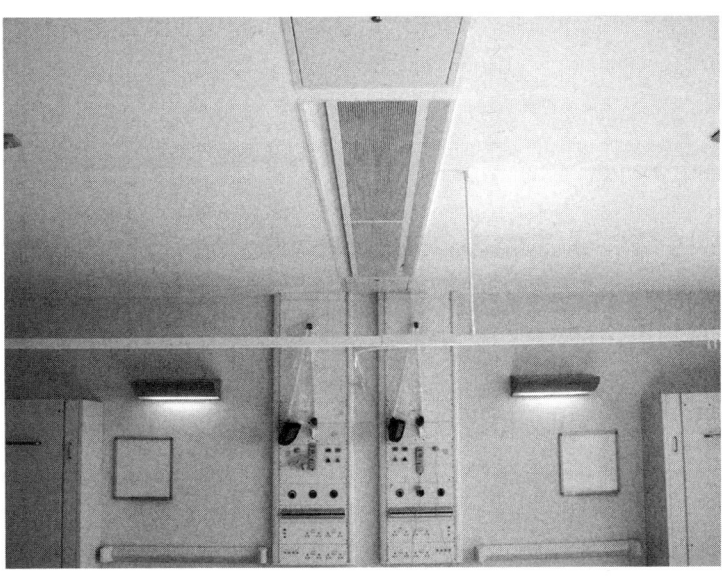

Figure 4–10 Chilled beam use in a hospital. *TROX Technik*

LABORATORY APPLICATIONS

Laboratories are extensive energy users and are considered the gas guzzlers of buildings. Standard labs consume about five to ten times the energy of standard office buildings, and clean rooms and labs with high process loads can consume up to 100 times more energy for the same square footage than a commercial building.[4] The extensive energy use comes from the heating and cooling of a once-through air system. In almost all chemical and biological labs the air system is 100 percent outside air. The air change per hour (ACPH) rate is also higher for laboratories than in office buildings. The ACPH rate for labs has been 10 to 15, whereas office buildings use only 6 ACPH. The driver of the once-through requirement is health and safety—avoiding the cross-contamination between labs and the increasing of contaminants levels (PPM). The purpose of a high ACPH rate is to dilute the contaminant levels. The high ACPH rate of 10 to 12 was adopted almost universally irrespective of the type of lab and operations. It is only recently with awareness of energy and sustainability, professional organizations such as Labs21 and others are encouraging design engineers to re-think the universal application of this standard. Design engineers, lab stakeholders, architects, scientists, lab technicians, and health and safety directors are encouraged to discuss the functions and use of their lab and arrive at the optimum air change rate. This calls for an integrated design approach and to establish an ACPH rate that is not too high or too low, and that will appropriately provide the necessary protection to lab users.

---

[4] "Laboratories for the 21st century: An Introduction to Low Energy Design," Labs21, http://epa.gov/lab21gov/pdf/lowenergy_508.pdf.

The extensive energy use in labs is from:

1. Heating and cooling of once-through air flow
2. 24/7 use of air
3. High process loads in labs
4. Cyclical and variable process loads
5. Reheat of air to meet thermostat settings
6. Fume hood exhaust requirements

Chilled beams reduce the energy consumption for five out of the six reasons for energy use. Both passive and active or a combination of active and passive chilled beams can be used in labs. The most common are the active chilled beams which use the minimum air flow required in the lab to perform further cooling of process loads. When the process loads exceed the capacity of the primary air and the chilled beam, then additional passive chilled beams can be added. The type depends on the performance of the beams and the setpoints of the labs.

The once-through system: The once-through systems in the labs cannot be avoided, for reasons of safety and health. However, this ventilation requirement can be put to use in a chilled beam application. Active chilled beams can use this required air flow into the lab to induce more air flow for the cooling capacity of the lab. Use of chilled beams will reduce the quantity of once-through air. Reduction in air flow will reduce the size of fans, air-handling equipment, fan energy use, duct sizes, shaft sizes, ceilings spaces, and heating and cooling energy costs.

High process loads: Chilled beams are cooling devices in the space, and they offset the process heat loads in the space. They reduce the air required from the central air systems for the process load that is offset by the chilled beams. Process heat from the space is carried away by water rather than air. This avoids heating and cooling the air from the outdoors.

Reheat of air and variable process loads: In all lab applications, reheat of cooling air is standard design. Reheating is required to meet the varying process loads. In conventional systems, at first the VAV system reduces the air flow; however, this is not sufficient in most applications because reheat kicks in. Reheat and precooling costs can account for as much as 20 percent of the costs of lab systems.[5] Chilled beams reduce the need for excess air, thereby reducing the reheat and precooling process.

---

[5] "Laboratories for the 21st Century: Best Practice Guide. Chilled Beams in Laboratories: Key Strategies to Ensure Effective Design, Construction, and Operation," Labs21, www.labs21century.gov/pdf/bp_chilled-beam_508.pdf.

Figure 4–11 Chilled beams in labs. *Asif Syed*

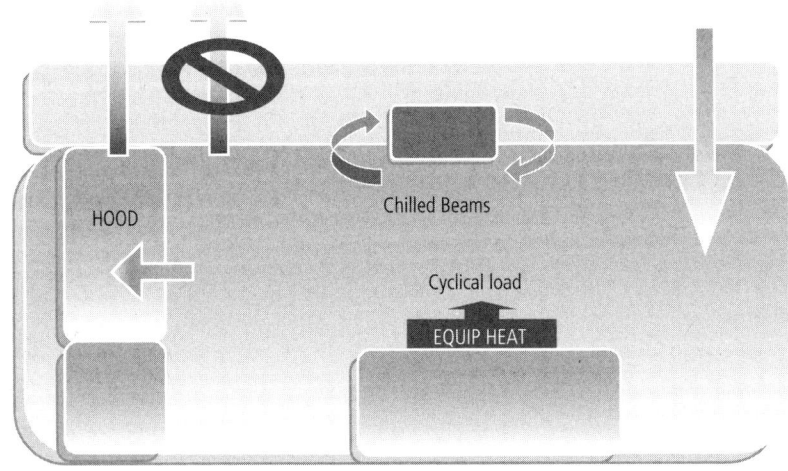

Fume hood exhaust: In labs that have large quantities of fume hoods, such as organic chemistry labs, the level of ventilation air is driven by hood requirements for safety. In such labs, chilled beams cannot offer much benefit. Chilled beams cannot help here, as minimum air flow is required to safely operate the fume hood. Other air flow reduction by using low-flow hoods and adjusting sash heights and air velocities, as safety strategies, must be considered.

CHAPTER 5

Underfloor Air Distribution (UFAD)

UNDERFLOOR AIR DISTRIBUTION IS A METHOD OF AIR CONDITIONING THE SPACE by supplying the air from the floor, using natural buoyancy forces to lift it to the ceiling, as opposed to the conventional systems, which supply air from the ceiling down to the occupants, working against the natural forces of buoyancy. The advantages of the system include energy efficiency, thermal comfort, individual occupant control, flexibility for frequent office restructuring, better indoor air quality, and lower costs for churn fit-out. UFAD technology uses an air plenum under the floor to supply air from floor air outlets. An air plenum is made between the floor and the structural slab. This requires a raised floor plenum of sufficient depth to transport the air from the supply source to the air outlets. The plenum space is easily accessible and provides the same level of access as a 2×2 ceiling tile—without the necessity of climbing a ladder. The space under the raised floor, primarily created for air flow, is also used for the distribution of other services, such as electrical power wiring, telephone and information technology cabling, security cabling, and fire alarms. The use of a raised floor plenum for other services makes the system flexible to modifications, due to ease of access. Given the dependence of business on computers, networks, VOIP, and so on, and the fast-paced changes in technology, the demand for flexibility in data cabling is ever increasing. The environmental and energy benefits come from the operating temperatures, which are much higher than in conventional systems. The combination of environmental and energy benefits with flexibility is the main reason for the growing popularity of the systems. The technology is not new to buildings; from the 1950s on, it has been used in data centers or computer rooms. The driver then was flexible cable management and efficient cooling of high heat loads. Today the drivers are: energy conservation, the environment, and flexibility in managing other building services.

The LEED 2009 *Reference Guide for Green Building Design and Construction* cites UFAD as a way to achieve individual occupant controls.[1] From the 1970s, buildings have been designed with this technology in Europe and Asia, and have worked satisfactorily. In the United States, UFAD technology was introduced in the 1990s, and several buildings have been designed to incorporate it. Initially, designs in the United States had mixed results. Projects where the designers and contractors tapped into European experience and expertise had very successful outcomes. Projects where design and construction techniques were not adapted to the technology had poor results and generated bad press for the technology. However, the U.S. building industry has learned the lessons of implementing the new technology and lately has built high-profile successful projects. Initial construction costs associated with UFAD systems are slightly higher than those of conventional systems, but the overall life-cycle costs are far less. Three major new projects in New York City, with approximately 5 million square feet of construction, have been built in the last five years using UFAD technology. The New York Times headquarters building uses UFAD[2] for 800,000 square feet of office space. The One Bryant Park building uses UFAD in addition to thermal storage, high-performance envelope, and daylight harvesting[3] as steps toward achieving LEED Silver certification.[4] Another major corporation in New York (not named for confidentiality) used UFAD in their new headquarters building.

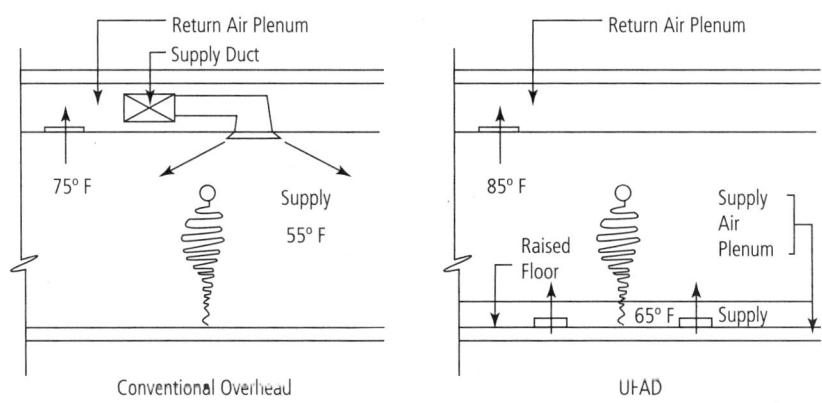

Figure 5–1 Conventional overhead air distribution and underfloor air distribution (UFAD). *Asif Syed*

---

[1] LEED *Reference Guide for Green Building Design and Construction*, 2009 edition.
[2] *New York Construction*, "NY Times Tower Newest 'Jewel' of NYC Skyline," December 2007.
[3] Bank of America and the Durst Organization Break Ground on the Bank of America Tower at One Bryant Park in New York City, 2004.
[4] Urban Green Council (New York chapter of the USGBC), database of New York City's LEED-certified buildings.

Benefits of underfloor air distribution

1. Energy efficiency: Reduced energy consumption in UFAD systems comes from:

   a. Higher air temperature: The supply air temperature of the UFAD system is higher than that of a conventional overhead air distribution system. The UFAD supply air temperature is 65°F, whereas conventional air systems use 55°F air. The benefit from higher temperature is twofold:

      i. The cooling system or the compressors will not have to work as hard, and will therefore use less energy. The refrigeration process of absorbing heat from the lower-temperature source (indoors) and rejecting heat to the high-temperature sink (outdoors) is more efficient when the source is at a higher temperature.

      ii. Economizer or free cooling is used in the air conditioning system. This is a method that uses outdoor ambient cooler conditions to cool an indoor heat-generating space. Office interior spaces require cooling even in the winter, when outdoor temperatures are cold. It is only the perimeter building envelope space that requires heat, as a result of skin loss. For UFAD systems, the economizer period can be extended. In conventional systems, economizer or free cooling stops when the outdoor temperature reaches 55°F. In UFAD systems, the free cooling can continue up to 65°F (with acceptable relative humidity).

   b. Thermal stratification: UFAD systems stratify the space above the head level up to the underside of the ceiling with higher temperatures (80° to 85°F). From the floor to just above head level is considered the occupied zone and is maintained at comfort temperature (72 to 78°F). Conventional systems maintain a uniform comfort temperature (72 to 78°F) over the entire height of the occupied space. In a space with 10-foot ceilings, about 6 feet is the occupied zone, and 4 feet is the unoccupied zone. Thus, UFAD systems condition only 60 percent of the volume of space to human comfort, and the remaining 40 percent is above the comfort zone. This process of stratification permits most of the heat from the lights, some of the solar heat gain from glazing, and a portion of the envelope heat gain to be returned directly to the cooling system, without any impact on the occupied zone of the space.

   c. Low fan energy: The air quantity required is the same for both conventional and UFAD systems. However, UFAD systems distribute air through a plenum with less air pressure loss compared to a conventional system with ductwork and variable air volume (VAV) boxes (which are dampers or valves or gates to control flow). The lower pressure loss results in lower fan energy consumption. Some conventional systems employ fans in the ceilings, such as in the fan-powered VAV systems; the energy consumed by these fans is also eliminated.

d. Diurnal temperature advantage: Diurnal temperature is the daily fluctuation in temperature between night and day. The lower temperature in the nights, when the building is not occupied, can be used to cool the building mass, for example the floor slabs. The air is directly in contact with the mass of the structural floor slabs. Conventional systems do not permit this. A night purge cycle can cool the building and increase the thermal inertia, which can be used during the daytime, reducing energy use and peak demand.

2. Flexibility, frequent office restructuring and churn costs

   Most office buildings have a very high churn rate. Churn is defined as the percentage of fit-out floor area that has to be re-fitted-out because a tenant or user has left the building, or because the needs of the tenant or user have changed. This is a very important matrix in a building's ongoing costs. It varies among buildings, but for some buildings or companies it can be very high. It is not unusual for an owner to change a space three or four times within ten years, because of varying business needs. In this situation, usually the space is gutted out, replacing ceilings, carpets, HVAC, electrical, and fire protection. UFAD offers the ultimate flexible system for high churn at lower costs. The raised floor of the UFAD system is built with 2-foot × 2-foot modular floor tiles on pedestals. This provides ultimate flexibility to access the space below the floor. Reconfiguration of office layout becomes easy for:

   a. Space planning: Space planning and interior office layout sometimes fall victim to the existing mechanical, electrical, and plumbing (MEP) infrastructure. The workstations are placed where there is power, or where columns are available to drop power. Slab poke-through is sometimes not possible, if there is a tenant in the floor below whose space cannot be disrupted.

   b. HVAC: A conventional system reconfiguration involves removal of ductwork, VAV boxes, ceiling tiles, and air outlets. Extensive effort is required to redesign and construct the space. Almost none of the material removed is ever reused. In a UFAD system, every cubicle or private office is provided with a floor outlet. In a reconfiguration of the floor plan, the floor air outlets can be easily removed and relocated along with the access floor tile.

   c. Electrical power wiring and low-voltage cabling: In a conventional system, electrical conduits and junction boxes are installed in the ceiling. The low-voltage cabling is in a cable tray or on J-hooks. Both electrical power and low-voltage cabling are brought to the work desk within walls or columns, through wiring or a power pole, or poked through from the floor slab, all of which are inflexible. When the work area is reconfigured, the changes are disruptive, expensive (due to the extensive field labor required), and wasteful (demolished material, wall outlets, and wiring are seldom reused). The flexibility requirements for low-voltage cabling are very high, especially with the fast-paced changes in technology for computer and telephone systems. UFAD systems minimize the disruption—work is limited to under the raised floor, electrical devices' floor outlets are simply relocated (rewired only if the BX wire does not extend to the new location), and ceilings are untouched.

d. Plug and play: An increase in occupant density or a revision of the spatial plan can be accommodated easily with extensions from the plug-and-play modular devices in the raised floor.

e. Electrified furniture: With UFAD, powered or electrified furniture is not required. At every workstation, a floor outlet can be provided. When electrified furniture systems have to be reconfigured, the wiring in the furniture has to be changed—another cost that is avoided with UFAD.

3. Thermal comfort and individual occupant control

Hot and cold spots are the top complaints in office buildings. The complaints are due to drafts and inadequate thermostats or zones of control. Both drafts and inadequate zones are inherent to conventional overhead design. Standard office buildings use about 1,600 to 2,000 square feet per thermostat or VAV zone. More zones can be provided, but the costs are high due to the need for more VAV boxes, controls, low-voltage control cabling, electrical wiring to VAV boxes, piping connections, or electrical power for heating coils. The primary reason that the VAV zone is limited to 1,600 to 2,000 square feet is that a larger VAV box creates excessive noise. If possible, buildings would increase the zone sizes to greater than 2,000 square feet to reduce costs; it is the noise that limits zone size, not occupant control. VAV systems vary the amount of air in the outlets; the turndown can be as low as 30 percent of the maximum. Ceiling air diffusers are fixed aerodynamic airfoils, and their performance changes when the flow changes. This leads to drafts at higher air flows and dumping at lower air flows. UFAD systems are free from drafts; because velocity at the floor outlet is low, the air outlet can easily be relocated to the occupant's preferred location. UFAD systems provide the ultimate in occupant thermal control; every occupant gets an air outlet, which he or she can easily adjust for the desired air flow.

4. Indoor air environment or quality

Contaminant control and ventilation effectiveness are the two benefits of UFAD systems for indoor air environments. Both benefits highlight the difference of the UFAD system from the inefficient conventional air distribution system, which is also known as a mixed-air system. In a mixed-air system, cold air is supplied to the space from the ceiling and returned from the ceiling. The warm air in the room, which is found around heat-generating objects, is "mixed" with cold air to form a uniform temperature throughout the room. Therefore the name "mixed air system."

a. Contaminants levels: In this mixing process, the contaminants—which are usually generated at the floor level, from copy machines or occupants—are also mixed up in the churning movement of the air. This increases the time contaminants stay in the space, increasing their density or parts per million count. In UFAD systems, the air flow is always steadily upward. The steady upward motion carries the contaminants to the exhaust grille, without increasing their density.

b. Ventilation effectiveness: Ventilation effectiveness is defined as the effectiveness of an air distribution system in removing internally generated pollutants from the ventilated space.[5] UFAD systems are more effective than conventional systems, and carry a higher rating.[6] Higher ventilation effectiveness means that the air is cleaner at the breathing zone than in the return air stream.

c. Noise levels: The noise levels of UFAD systems are low because of lower velocity through air outlets. The only source of noise is from underfloor air pathways. Maintaining these low velocities may make UFAD systems too quiet for some very low occupancy offices that need sound-masking systems.[7] For standard offices, sound-masking systems are not required. The noise from the conventional systems mostly comes from diffusers, air velocity in ducts, vibrations of fans transmitted through ducts, whistling caused from volume dampers, and VAV boxes (which are just automated volume dampers). All the noise-generating devises are eliminated in UFAD systems.

TABLE 5-1 VENTILATION EFFECTIVENESS OF DISPLACEMENT AND CEILING DIFFUSERS

	Ventilation Effectiveness	
Measurement plane	UFAD	Ceiling Diffuser
Seated	1.3 to 2.0	0.90 to 0.97

---

[5] Hazim Awbi, *Ventilation of Buildings*, 2nd ed., London: Spoon Press, 2003.
[6] Ventilation for Acceptable Indoor Air Quality, ASHRAE Standard 62.1-2007.
[7] Jack Geortner, "The Upside of Underfloor Air-Distribution Systems," *RSES Journal*, November 2005.

Figure 5–2 Tate Access Floors. *Image provided by Tate Access Floors, Inc.*

Figure 5–3 UFAD diffuser. *TROX Technik*

TABLE 5-2 DIFFERENCES BETWEEN CONVENTIONAL AND UFAD SYSTEMS

No.	UFAD	Conventional
1	Supply air temperature is 65°F, and return air temperature is 85°F.	Supply air temperature is 55°F, and return air temperature is 75°F.
2	Supply is from floor with a raised floor supply air plenum.	Supply is at ceiling with supply ducts in the ceiling.
3	Air outlets are floor devices.	Air outlets are standard ceiling diffusers, which are louver face, slots, or wall grilles.
4	Each workstation gets an air outlet in the floor, about 8 inches in diameter.	Ceiling diffusers generally serve multiple occupants and are generally 2 feet × 2 feet.
5	Lower noise level results from low air velocity through the air outlet.	Higher noise levels from higher air velocity through the outlet. Diffuser noise is one of the main sources of noise in the HVAC system.
6	Total air quantity flowing through the space is the same as in conventional systems.	Total air quantity flowing through the space is the same as in UFAD systems.
7	Quantity of ventilation or outside or fresh air can be reduced to provide the equivalent level of indoor air quality as a conventional system. This reduces energy consumption as outside air is hot and humid (summer) or cold and dry (winter), and energy is used to condition it to room comfort level.	For the same indoor air quality, higher amounts of outside or ventilation or fresh air are required. This increases the energy costs to cool the additional air quantity.
8	Cooling apparatus (air-handling units) is different from conventional. To cool and dehumidify air to 65°F, only a desired fraction of air is dehumidified by cooling, and then mixed with warmer room air to create 65°F air. This is an energy-efficient process, as dehumidifying the entire volume of air uses more energy.	Cooling apparatus is standard. Conventional units cool and dehumidify air to 55°F. Here the entire volume of air is dehumidified. In most applications, the extent of dehumidification is more than needed, but the system requires it to cool the air to maintain the desired temperature. This is an inefficient and energy-consuming process.
9	In a free cooling cycle, outside air up to 65°F can be used, giving more hours of free cooling. Free cooling is defined as a means of using the outside environment to cool the indoors.	In a free cooling cycle, outside air up to 55°F can be used, reducing the number of free cooling hours.
10	In UFAD systems, air is separated into two streams: one for dehumidification and cooling, and another for reheating. This saves energy by not overcooling or over-dehumidifying.	Entire supply air is cooled and dehumidified in one single process. This results in overcooling or over-dehumidification of air, a waste of energy.
11	Contaminant level (parts per million or PPM) is lower.	Higher PPM of contaminants.
12	No short circuit between supply and exhaust, providing extremely effective ventilation of the space.	Possibility of short circuit between supply and exhaust, reducing the effectiveness of ventilation.
13	No air drafts or cold air dumping.	Usually drafty environment or dumping of cold air due to reduced flow in VAV systems.
14	Uses less energy due to higher temperature supply (65°F).	Uses more energy due to lower temperature supply (55°F).
15	Additional energy benefit due to extended period of free cooling to 65°F.	Period of free cooling is limited to 55°F.
16	Computational fluid dynamics (CFD) analysis is required to validate design.	Does not require CFD analysis.

VALIDATION OF UFAD DESIGNS WITH CFD ANALYSIS

For conventional systems—such as VAV air distribution, which has been used for the last thirty years—the lessons learned have been translated into designs, but UFAD designs are relatively new in the United States. The designers are not as familiar with the details of its design and operations, and have limited experience with the technology. The best and most cost-effective way of validating a UFAD system is with computational fluid dynamics (CFD) analysis. The analysis permits verification of the design with a virtual mock-up. For UFAD design, the most recommended subject of CFD is the air flow velocity in the below-floor supply air plenum. Excessive velocity and pressure of air can lead to lack of uniform air distribution above the floor. The velocity and pressure depend on the quantity of air discharges into the plenum, their locations, the plenum depth, and the discharge velocity. All of these variables can be modeled in a CFD analysis very early in the design, in the early schematics. The depth of the plenum has an impact on the architecture of the building. Locating air discharges to achieve uniform pressure and velocity has an impact on air shaft or riser locations in

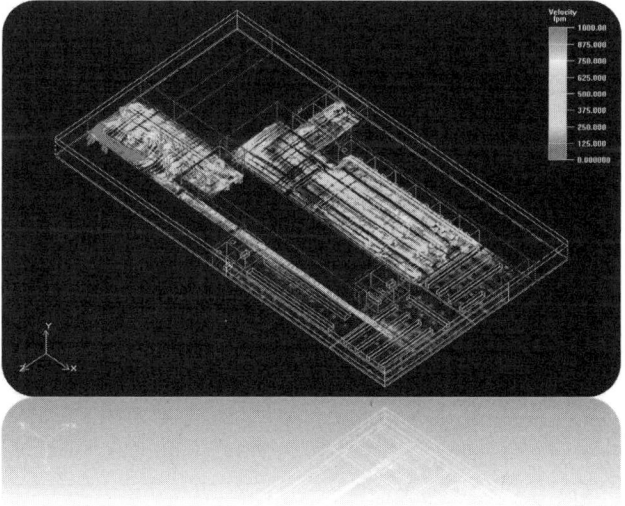

Figure 5–4 Incorrect air distribution under raised floor. Very high velocity streams of air prevent air flow through floor outlets.   *Asif Syed*

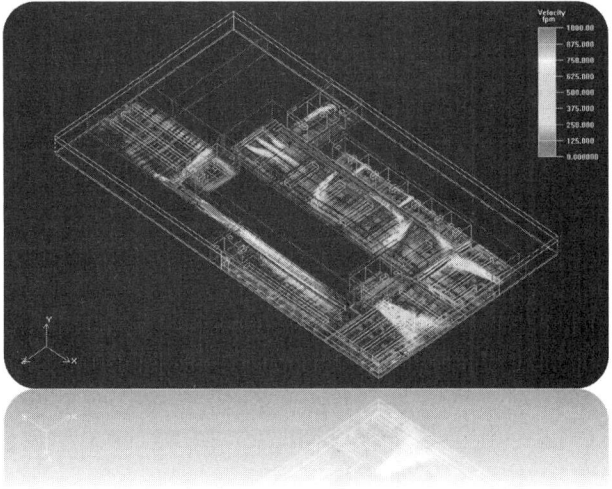

Figure 5–5 Correct air distribution under raised floor provides uniform low velocity throughout the plenum. *Asif Syed*

the core, a major architectural decision. Architects do not like to change the building core design after schematics; all input has to be given in pre-schematics. CFD analysis helps in firming up the architectural elements very early in the design process in the conceptual phase. UFAD design is an integrated design process, where synergies of different trades are combined together to maximize the benefits. By validating the design, CFD avoids costly mistakes. At one time, CFD analysis required powerful computers, but present-day personal computers can perform these analyses. The software for CFD analysis requires licensing and training of engineers, which is an additional cost burden to the design team. The CFD software can perform more analyses than just UFAD and has become essential in design of buildings.

COST OF UFAD SYSTEMS

The cost of UFAD systems varies from project to project, primarily based on the size of the project, the type of UFAD HVAC system chosen, the ratio of perimeter area to interior area, the experience of the cost estimators, and so forth. A total life-cycle cost for thirty to fifty years is the best way to judge the overhead and UFAD system cost. The costs specific to the project—the initial capital costs, operating energy costs, operating maintenance costs, and churn costs—must be estimated. Capital costs must be estimated by a cost estimator or construction manager who is familiar with the technology and has experience with UFAD systems. The energy cost estimate is best when a 365-day 24-hour energy model is used along with accurate utility rate information. The maintenance cost of the system must be estimated by a facilities management company with experience in maintaining UFAD systems. The churn rate must be developed by the landlord and user or the real estate agent for the space. For most UFAD systems, even with a low churn rate, the life-cycle costs are lower than those of a conventional system due to ongoing yearly lower energy costs. Generally the UFAD systems cost slightly more in first or capital costs than conventional overhead systems. One such study, done for the U.S. General Services Administration (GSA),[8] shows that the increase in construction cost can vary from $2 to $6 per square foot, depending on the type of UFAD system and the size of the floor plate. The results from the study are shown in Table 5–3.

Another study[9] shows that $1 of additional costs saves $5 over the building life cycle. A case study by Carnegie Mellon University[10] shows that the additional first costs of a UFAD system (including raised access floor) was $0.27 per square foot. The savings at the first churn with the UFAD system was $4.66 per square foot.

TABLE 5-3 INCREMENTAL COST OF UFAD SYSTEMS IN U.S. $ PER SQUARE FOOT

UFAD Type	Type A	Type B	Type C	Type D
20,000 sq. ft	$5.82	$2.60	$3.50	$3.49
35,000 sq. ft	$4.04	$2.18	$3.20	$2.90
50,000 sq. ft	$3.57	$1.57	$2.33	$2.28

(Reproduced from Webster et al., 2006)
Notes:
Baseline system: Overhead air distribution with VAV boxes and RH coils along the perimeter and standard ceiling diffusers.
Type A: All constant air volume (CAV) system for interior and perimeter spaces.
Type B: Constant air volume for interior spaces and variable air volume (VAV) for perimeter spaces.
Type C: All variable air volume (VAV) for interior and perimeter spaces.
Type D: All variable air volume (VAV) with modulating dampers.

---

[8] Tom Webster, Corinne Benedek, and Fred Bauman, *Underfloor Air Distribution (UFAD) Cost Study: Analysis of First Cost Tradeoffs in UFAD Systems*, The Center for the Built Environment, University of California, Berkeley, September 2006.
[9] Germershausen, M., 2000. "Wired for Success," *Buildings* 94, No. 7.
[10] Loftness, V. et al., "Sustainable Development Alternatives for Speculative Office Buildings: A Case Study of the Soffer Tech Office Building," Center for Building Performance and Diagnostics, Carnegie Mellon University School of Architecture, 1999.

MYTHS ABOUT UFAD SYSTEMS

1. Cold feet—Wrong. There is a general perception that since air is supplied at lower level near the floor, it leads to cold feet. This is not true, as UFAD air is supplied at 65°F, which is not cold. However, if the conventional system air is supplied at 55°F at a lower level near the floor, it is cold and causes discomfort.

2. Calculations are the same as for conventional systems—Wrong. Heating and cooling load calculation methodology is different for UFAD and overhead systems. Stratification of heat above the occupied zone must be taken into account. Except for a few, most cooling load calculations software packages do not have a methodology to capture the effect of stratification, higher supply air temperature, higher return air temperature, splitting of air stream for optimum dehumidification, and mixing of air streams for higher temperature. The engineers have to factor all these variables accurately to arrive at the right system performance.

3. UFAD systems require more air—Wrong. The total air required for both UFAD and conventional systems is the same. The UFAD systems do not require larger duct shafts and fans.

4. Mechanical systems are the same for both—Wrong. UFAD systems require different fan systems and configuration that permit the air to bypass the coil, but that filter it and mix it with colder dehumidified air to produce 65°F air. This configuration of the fans is not common and is not a standard offering by most manufacturers of HVAC systems. The units have to be custom or semi-custom designed.

5. Large ductwork is required for UFAD—Wrong. Duct sizes are the same for both conventional and UFAD systems. Some early installations which were not designed correctly had high air leakage. To compensate the air leakage, additional fans and ducts were retrofitted. A correctly designed UFAD system does not require larger duct or fan.

6. Higher floor-to-floor height is required for UFAD—Wrong. There is a perception that the height of the underfloor air plenum adds additional height between slab to slab. This is not true. Additional floor-to-floor height is not required, and building height does not increase. The ceiling plenum height is smaller by the floor plenum height, making the overall slab-to-slab height the same for both UFAD and conventional overhead systems. In concrete structure or flat slab construction the floor-to-floor height can even be less than that for conventional systems.

IMPACT ON BUILDINGS

FLOOR-TO-FLOOR HEIGHT

Steel structure buildings: In buildings with steel beams, the floor-to-floor height is the same for UFAD as for conventional overhead systems, or slightly smaller by about 3 to 4 inches. The ceiling plenum height is smaller by the floor plenum height, making the overall slab-to-slab height the same for both UFAD and conventional overhead systems. The primary function of the space above the ceiling is as a return air plenum. Air gathered from the ceiling grilles is returned to the air shaft through the plenum. The depth of the beam is important, because there has to be space below the bottom of the beam and the ceiling for air to pass. This height varies based on the floor plate size, return air shaft location, and quantity of shafts. For a standard office building, this number is about 6 to 8 inches, but this is not universal; the height has to be verified for project-specific conditions. In conventional systems, the depth required for VAV boxes and ductwork is about 20 inches. This 20-inch depth is transferred to the raised floor plenum.

Concrete structure buildings: In concrete buildings, higher floor-to-floor ceilings can be achieved with UFAD. In buildings with concrete structure, the concrete slab with flat slab or shallower beams results in lower floor-to-floor heights. Shallower beams combined with UFAD give a greater reduction in floor-to-floor heights than steel buildings.

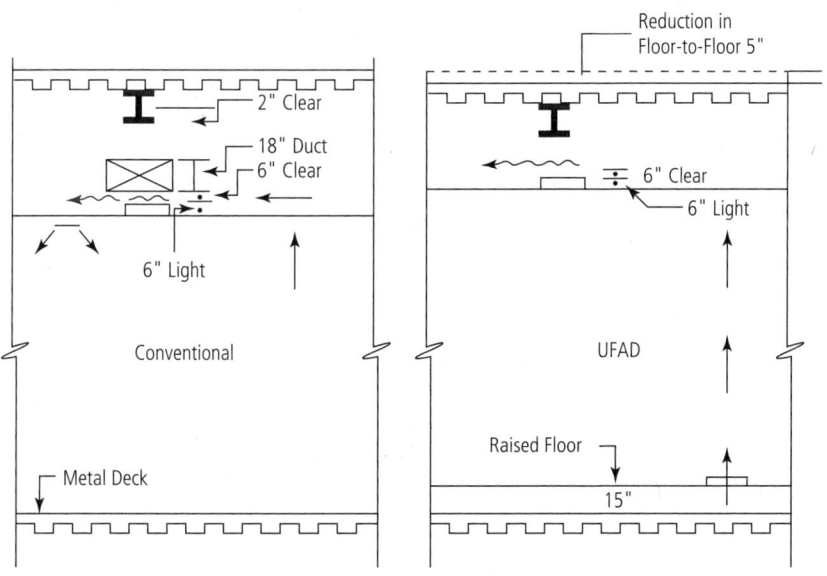

Figure 5–6 Floor-to-floor heights of UFAD and conventional systems. *Asif Syed*

Figure 5–7 Floor-to-floor height savings with concrete flat slab or concrete slab with shallow beams.   *Asif Syed*

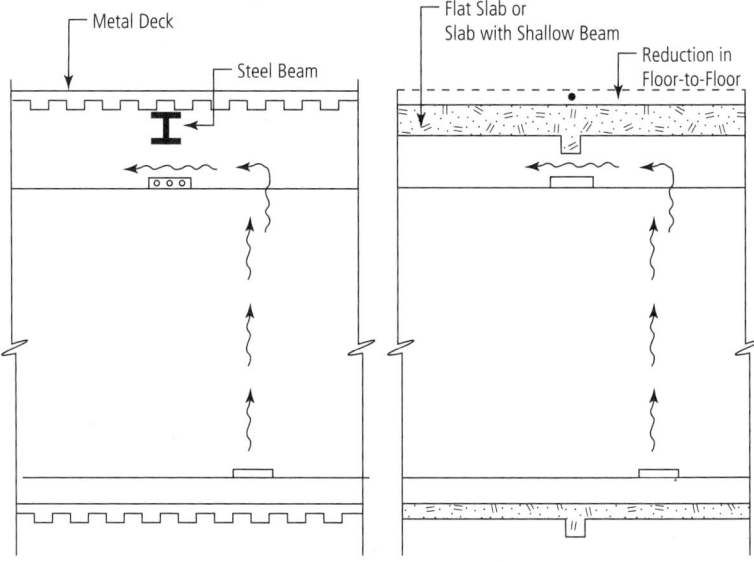

UFAD IMPACT ON BUILDING CORE SPACES

In UFAD projects, the buildings' core spaces, such as toilets, electrical rooms, telecommunication rooms, stairs, and elevator stops, have to be coordinated with the UFAD plenum height. It is important to establish the plenum height in early phases of design. Figures 5–8, 5–9, and 5–10 provide details of this. The toilet fixtures are wall-mounted, due to solid lightweight fill. Alternatively, the metal deck can be used at the toilet areas, which are usually located at the core of the building. The metal deck can be raised up to the level of the raised floor. This permits toilet fixtures to be either wall- or floor-mounted.

CRITICAL ISSUES OF UFAD DESIGN

Air leakage or plenum Integrity: Air leakage is the primary issue that has brought bad press to UFAD designs. Excessive air leaks, from air plenums to the exterior at the curtain wall and to core elements, reduced the air flow to the occupied spaces, leading to discomfort. However, the construction industry has learned this lesson and has emphasized the importance of this issue by developing details for joints under the plenum and leakage factors for air capacity. Tate Access Floors, Inc. and the Construction Specifications Institute (CSI)[11] recommend a 10 percent leakage factor for air losses to outside the occupied spaces

---

[11] *Architect's Guide for Detailing & Specifying Access Floor Air Plenums*, Tate Access Floors, Inc., 2009.

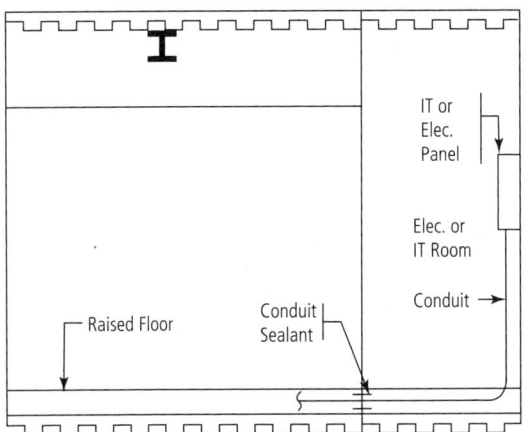

Figure 5–8 Electrical or information technology rooms in UFAD. *Asif Syed*

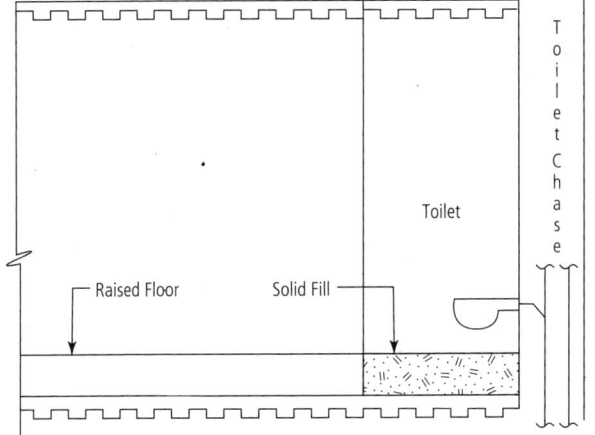

Figure 5–9 Toilet section in UFAD systems. *Asif Syed*

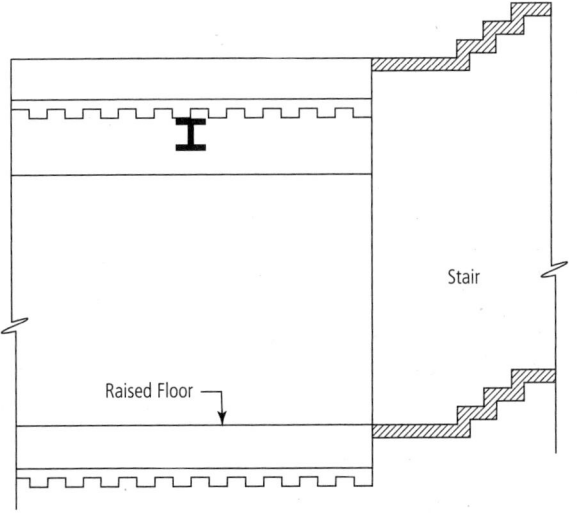

Figure 5–10 Stair section in UFAD systems. *Asif Syed*

Figure 5–11 The New York Times headquarters building uses UFAD. *Asif Syed*

(referred to as category 1) and an additional 10 percent air loss factor for leakage into the occupied space (referred to as category 2). Engineers are used to designing air systems with the assumption that air stays in a sealed duct, and architects designing joints between walls and floors do not anticipate the need to contain or seal air. Both assumptions have to be rethought with UFAD design. UFAD requires that the wall joints, and joints between wall and under the floor, be airtight, and that the air leakage factor of the fan systems be higher than that of sealed ducts.

Integrated design: UFAD designs are interdisciplinary. The plenum designed by the architect is used by the mechanical engineer to supply air. The underfloor air plenum is built by craftsmen who are not used to building air-tight plenum as part of their trade. This calls for an integrated design where all team members understand the system and buy into the concept.

Mechanical systems component design: It is important for the mechanical engineer to become familiar with the design of the UFAD system. The system

Figure 5–12 One Bryant Park uses UFAD. *Asif Syed*

components and configuration is different from the conventional air system. The air-handling unit configuration includes a bypass around a cooling coil. The plenum design is important for the performance of the UFAD. The air velocities in the plenum have to be low, because high velocities will lead to lack of air flow through the diffusers. This is a difficult problem to solve after the installation, therefore the plenum design has to be carefully designed with appropriate points of air entry, and validated by a Computational Fluid Dynamics software.

CHAPTER 6

Displacement Induction Units (DIU)

DISPLACEMENT INDUCTION UNITS (DIU) ARE UNIQUE PIECES OF EQUIPMENT in the air conditioning system. They are similar to under-the-window-sill fan coil units and induction units, but their performance is uniquely superior to other similar systems. The DIUs work on the principle of displacement ventilation. DIUs have a water coil in the space that is to be conditioned and are therefore local cooling units, unlike conventional central air systems that have a water coil located remotely and air supplied and returned with large duct and fan systems. DIU makes possible displacement ventilation with locally cooled units and this is what makes them unique. They combine the benefits of displacement ventilation and local cooling. The units use water as the cooling and heat transfer medium. Water is more efficient than air for transporting heat, which makes the system more energy efficient than conventional (all-air) systems. Additionally, the DIUs have the benefits from the displacement ventilation, which is energy efficient, better thermal comfort, smaller ducts, smaller fans, and better indoor air quality. Displacement induction units are similar to chilled beams and operate on the same principle, however, chilled beams are located in the ceilings and DIUs are located at floor level.

The ventilation air required for occupants in the space is the driver for the air movement through the room, eliminating the need for fans. As with chilled beams, the induction nozzles induce air flow over a cooling coil, which cools the air. Because the units move air with a displacement methodology, they are located at floor level. The units are generally located under the window sill; DIUs are 11 to 15 inches wide.

The units are similar to the induction units that were popular in the 1960s and 1970s in office buildings, at the perimeter. One example was the use of induction units for the twin towers of the World Trade Center, where the 12 feet of the perimeter were serviced by these units. The main difference between induction units and DIUs is that induction units discharge air upward toward the glass and DIUs discharge air toward the floor. This makes them suitable for displacement ventilation. Additionally the induction technology of the present day has made significant improvements over the 1960s and 1970s equipment. DIUs operate at much lower noise levels and static pressure, making them quieter and suitable for even the quietest applications such as classrooms.

In the DUI system, the only air required from the air-handling unit is the ventilation or fresh air to meet the ventilation requirements of the occupants. The ventilation air required is only about one-third to the air required in a conventional system. The major benefit of DIU comes from its local cooling methodology, which reduces the fan energy consumption. In an all-air central variable air volume (VAV) system, air is cooled at a remotely located mechanical room and transported to the space via large ducts. Fan energy is a big component of the HVAC system energy, and reduction of this requirement contributes to the overall reduction in energy consumption. In DIU, a local cooling system, water is cooled remotely and transported to the local cooling device. Pumping water is more efficient than moving air. Water has the capacity to carry more heat content than air. The DIU is a local cooling device, premanufactured and readily available for displacement ventilation in a variety of applications. There are multiple manufacturers of DIUs, which provides flexibility to architects and engineers to choose from different dimensions suitable for their project needs. The displacement induction unit works on the principle of inducing air flow in the room, with the motive force of primary air. The primary air is the ventilation air or outside or fresh air that is required to meet the ventilation requirements of the occupants. The secondary air is the air that recirculates in the room. The primary air is usually one-third, and secondary air is usually two-thirds. The two-thirds circulated air moves across a cooling or heating coil, producing the cooling or heating locally within the space.

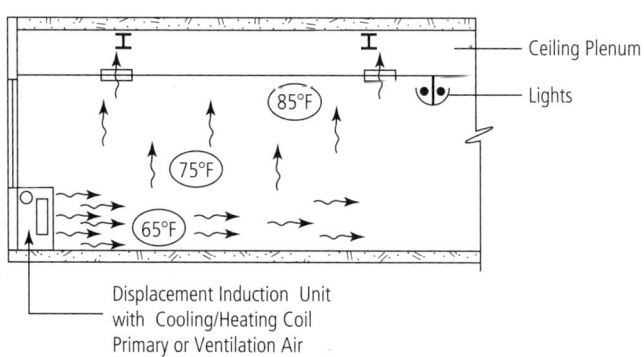

Figure 6–1 Displacement induction principle. *Asif Syed*

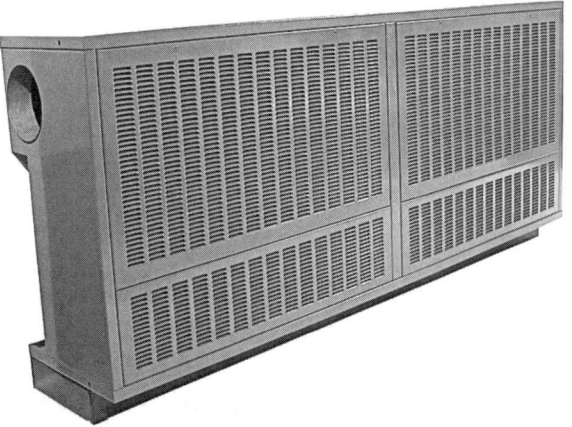

Figure 6-2 Trox Displacement Induction Unit. *TROX Technik*

The reduction of the central system capacity by two-thirds reduces the size of air-handlers and ducts, thus reducing costs, relieving congested ceiling spaces, and increasing ceiling heights. The fanless DIUs eliminate the cost of electric wiring, which is a relatively high cost for a comparative unit due to the extensive field labor involved. The DIU does not have any moving parts and therefore has very low maintenance. The only maintenance involved is the cleaning of nozzles, usually with a vacuum cleaner for dust accumulation.

BENEFITS OF DISPLACEMENT INDUCTION UNITS

1. Low energy consumption
2. Thermal comfort
3. Lower noise levels
4. Space savings
5. Improved indoor environment
6. Lower electrical costs
7. Lower maintenance costs

LOW ENERGY CONSUMPTION

1. Supply air temperature is higher by 10°F. The DIU can use 65°F air, whereas conventional air systems use 55°F.
2. The compressors don't have to work as hard and therefore use less energy. The refrigeration is operating at a lower temperature differential between source (indoors) and sink (outdoors), making it more efficient.

3. The higher-temperature air means that more economizer or free cooling is available.
4. Displacement systems stratify the space above the occupied level (usually above the head of a standing person) to the underside of the ceiling with higher temperatures (80 to 85°F). This reduces the percent of conditioned space, lowering the energy used to operate the system. In displacement systems, a portion of heat from lights and equipment such as computers, is vented directly to return air, reducing the load of the occupied space.
5. Passive operation of the unit is possible in heating mode. In climates that require heating and cooling, the same unit works in both heating and cooling mode. In unoccupied periods of the day, heating still has to continue to maintain a night setback temperature, to avoid freezing of the building. During this time the unit operates without primary air. Conventional comparable systems such as fan coils or central VAV systems have to operate fans that consume energy.

THERMAL COMFORT

1. Displacement ventilation systems are low velocity systems and operate on the principle of natural buoyancy forces and therefore eliminate drafts from the environment. Most conventional systems cause some form of draft and dumping of cold air, irrespective of how carefully they are designed. Hot and cold spots are the top complaints in buildings. The complaints are due to drafts and inadequate thermostats or zones of control. Displacement induction units provide individual zone control, and each unit can be provided with its own thermostat. In large common spaces such as classrooms, multiple units can be controlled with one thermostat.

LOWER NOISE LEVELS

1. The noise levels of displacement systems are low due to lower velocity through air outlets. The velocity of a displacement air outlet is about seven to ten times slower than that of conventional systems. High air velocity is the major source of noise in conventional systems.
2. Displacement induction units are cooling devices without a fan, which makes them quieter than fan coil units. This benefit of DIUs makes a big difference in the design of teaching environments, classrooms, and hospital patient rooms.

SPACE SAVINGS

1. Displacement induction units are local cooling devices that use water as the medium to transport heat. Water carries more heat for less volumetric flow, allowing for smaller ducts. Duct sizes are reduced by two-thirds, reducing the size of the shafts and ceiling plenum spaces.

IMPROVED INDOOR ENVIRONMENT

The indoor environment is improved by contaminant control and ventilation effectiveness in displacement systems. Both benefits are derived from the physics of their operating principle. Conventional air distribution systems (also known as mixed-air systems) lack these due to their air delivery methodology.

1. Contaminants generated at the floor level, from copy machines or occupants, are directly raised up to the ceiling and into the return air grilles, along with the displaced air. Displacement systems limit the churning movement of the air, reducing the volumetric space for contaminants and build-up of contaminants decreasing their density or parts per million counts. In displacement systems, the air flow is always steadily upward. The steady upward motion carries the contaminants to the exhaust grille, without increasing their density.

2. Ventilation effectiveness is defined as the effectiveness of an air distribution system in removing internally generated pollutants from the ventilated space. Displacement systems are more effective than conventional systems and carry a higher rating.

LOWER ELECTRICAL COSTS

1. Displacement induction units do not have a fan to circulate air. Not having a fan reduces electrical wiring costs, which are mostly from field labor and can be expensive. It is important when making a comparative cost between DIUs and other systems such as fan coils and VAV boxes, that the cost of electrical power connections be included to accurately represent total costs, not just mechanical costs.

LOWER MAINTENANCE

1. Displacement induction units do not have any moving parts such as fans and motors. The only moving part is the control valve in the piping. The control valve opens and closes to maintain temperature. Lack of mechanical moving

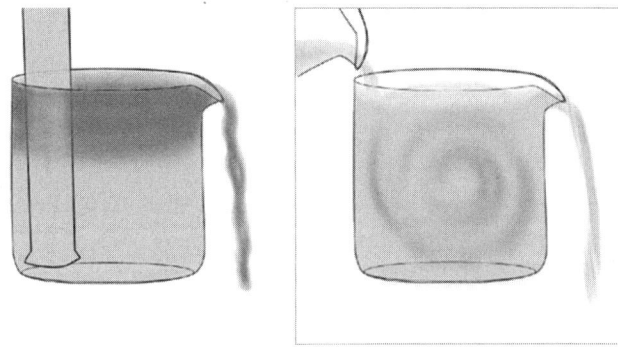

Figure 6–3 Displacement induction contaminant control. *XeteX, Inc. Copyright 2011*

parts reduces the maintenance on the units. The maintenance on the units is limited to visual inspection and cleaning of the unit and the induction nozzles, which is usually preformed with a vacuum cleaner, cleaning of the wire mesh filter in front of the cooling and heating coil, and cloth wiping to remove dust.

HISTORY OF INDUCTION UNITS

Though displacement induction units are relatively new, their operating principle of induction is not new to the building industry. Building HVAC operators are familiar with the technology, as standard induction units were very popular in buildings in the 1960s and 1970s, both for high-rise construction and low-rise construction. Induction units were used to condition the perimeter spaces of the buildings. The advantages of the units included smaller ducts. In high-rise buildings, smaller ducts significantly increased the floor or useable space efficiency by reducing the size of the air shafts; all-air systems would have added large shafts. The units also decoupled the perimeter from the interior spaces, thereby providing a means of heating the perimeter spaces, while the interiors required cooling. Most of the high-rise office buildings in New York, Chicago, and Boston had them. The World Trade Center in New York had 24,000 induction units in both towers.[1] The induction units of the 1960s and 1970 had high noise levels due to high-pressure primary or driving air. Modern day induction units have made significant design and performance improvements. They are much quieter, as they use low-pressure air. Induction units went out of fashion with the advances in the façade performance with double wall windows, low E coating, thermal breaks in window frames, and insulation. In most applications

---

[1] *Engineering News-Record*, 1967. April 13, 1967, RECORD HVAC CONTRACT LET FOR WORLD TRADE CENTER, http://911research.wtc7.net/mirrors/guardian2/wtc/eng-news-record.htm.

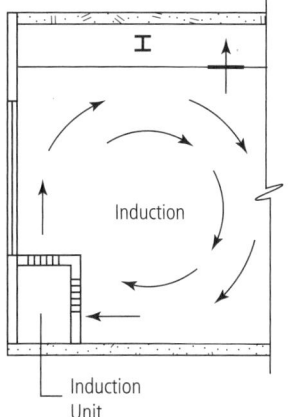 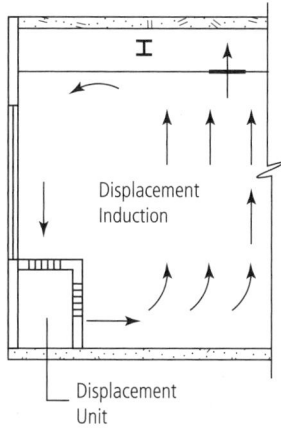

Figure 6–4 The difference between induction units and displacement induction units. *Asif Syed*

with the present-day wall performance, a heating element under the window is not required and air can be distributed from ceilings. The real estate industry was quite happy to gain 15 to 18 inches of space on the perimeter of buildings—valuable rentable floor space.

THE DIFFERENCE BETWEEN INDUCTION UNITS AND DISPLACEMENT INDUCTION UNITS

The basic difference between the two is the way they discharge the air. Standard induction units work on the conventional mixed-air principle, whereas displacement induction units operate on the displacement principle. Standard induction units discharge air from the top (window sill) and maintain flow along the window. Displacement induction units discharge air at floor level, making them suitable for displacement ventilation. Displacement induction units discharge air at a low velocity suitable to the displacement air flow principle. Standard induction units discharge air at a higher velocity to provide the mixing of air streams in the mixed-air system.

APPLICATIONS

1. The teaching environment: classrooms
2. Health care and hospitals: patient rooms
3. Perimeter buildings
4. Operable window spaces

THE TEACHING ENVIRONMENT (CLASSROOMS)

Displacement induction units can solve all the HVAC problems in a classroom: excessive noise, poor ventilation effectiveness, cold drafts from dumping of cold air, and poor air quality or contaminant control. In addition to this, displacement induction units have the advantage of local cooling, which saves fan energy used in the conventional systems.

Displacement induction units are installed along the perimeter under the window sill. The units are about 11 to 15 inches deep and take up the entire length of the classroom. Early coordination with the structural engineer and architect will eliminate the conflict between perimeter structural steel or concrete columns and induction units, duct penetrations, or pipe sleeves. The duct penetrations are usually small, generally about 6 to 8 inches in diameter. The ductwork is generally installed in the ceiling of the floor below. The ductwork can also be installed vertically along the perimeter wall. There are different configurations available for this, but it is important to note that the ductwork is only one-third that of a conventional system and therefore small.

Air is discharged from the DIUs at very low velocity. The low-velocity air flow eliminates noise. Problems with noise are caused by the fans in the fan coil system and high air flow from the all-air VAV system. DIUs do not have another problem commonly encountered in classrooms with a VAV system—dumping of cold air and breezy drafts. The drafts and dumping happen due to reduced air flow at low heat loads; diffusers designed for peak summer loads have to perform at low air flows at non-peak times leading to cold drafts. The displacement air movement has no cold drafts common in variable air volume systems.

The displacement air movement increases ventilation effectiveness and reduces contaminant levels. Ventilation effectiveness is defined as the effectiveness of an air distribution system in removing internally generated pollutants from the ventilated space. Displacement systems are more effective than conventional systems and carry a higher rating.[2] For a classroom where most of the occupants are seated for most

TABLE 6-1 VENTILATION EFFECTIVENESS OF DISPLACEMENT AND CEILING DIFFUSERS

	Ventilation Effectiveness	
Measurement plane	Displacement	Ceiling Diffuser
Seated	1.3 to 1.95	0.90 to 0.97

---

[2] Sean Badenhorst, "Under Floor Air Distribution," August 2002.

of the time, the ventilation effectiveness is higher. Higher ventilation effectiveness means that the air is cleaner at the breathing zone than in the return air stream.

DIUs are especially well suited for retrofit of the classrooms. Most classrooms built in the 1950s to 1970s did not have air conditioning. The classrooms were just not used in summer. It was not the standard to air condition the classrooms both in K–12 and in higher education buildings. With the high rate of growth in higher education and the need for classes in summer, several older classroom buildings are retrofitted. Older classroom buildings have heating only with cast iron radiators under the window sill. This presents an opportunity to retrofit them with air conditioning with displacement units, with all its benefits. Several classroom buildings have been

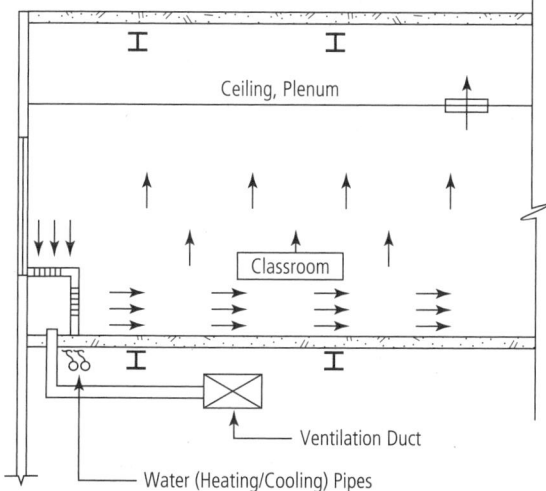

Figure 6–5 Displacement induction duct and piping arrangement. *Asif Syed*

Figure 6–6 Displacement induction units at St. John's University, D'Angelo Center. *Asif Syed*

Figure 6–7 Interior classroom with induction units, St. John's University, D'Angelo Center.
Asif Syed

Figure 6–8 Retrofit of fan coil system with displacement induction units, Ocean County Community College.
TROX Technik

Figure 6–9 Retrofit of cast-iron radiators with DIUs at St. John's University, St. John's Hall.
Asif Syed

Figure 6–10 Displacement induction units in classroom building. *Asif Syed*

retrofitted with fan coils, due to the lack of familiarity of DIUs in the design and construction community. The fan coils are noisy and increase their noise levels over time due to deterioration of fan rotating shafts. The retrofit with DIUs generally involves installing a central fresh air-handing unit and ductwork distribution. The ductwork required is only one-third of that required for central all-air systems such as VAV. There are several buildings which have fan coil units or classroom package terminal units. When these buildings are up for renovation, they can be retrofitted with DIUs. This will significantly improve the performance of the classroom and reduce energy consumption and improve acoustics.

HEALTH CARE: PATIENT ROOMS (NEW HOSPITAL)

In hospitals, DIUs can be used in patient rooms. The common practice in the healthcare designs are all central air VAV systems. Patient rooms can be provided with either central air or local displacement induction units. DIUs are better suited for patient rooms because they provide better energy efficiency, better indoor air quality, less noise, and smaller central duct systems. The air required for ventilation is used to for induction through the DIU and then exhausted through the toilet. For new hospitals, displacement induction units present a great opportunity to reduce energy costs. The fan energy from the central fan systems, the displacement ventilation energy savings, and more free cooling hours can add up to savings in hospitals, which are energy-intensive buildings. The average energy use of a hospital can be six to eight times that of a standard office building. The reduction in the size of large air handlers saves capital costs and relieves congestion in ceiling spaces by using smaller ducts.

Hospital corridor ceilings are the only pathway from central fan rooms to the patient rooms. Hospital designers know how congested the ceiling spaces in the corridors are, because of large duct sizes. This leads to lower ceilings or higher slab-to-slab area for the entire floor, to accommodate for space in the corridor. As a result, there can be difficulty in installation; most construction change orders for ductwork are due to the inability to accommodate all the ductwork in the ceiling. With a DIU system, supply duct sizes can be reduced by two-thirds that of all-air VAV systems.

HEALTH CARE: PATIENT ROOMS (EXISTING HOSPITAL RENOVATIONS)

Induction units were commonly used in hospitals built from the 1960s to the late 1970s. Most of these hospitals are up for renovation, as the installations are close to thirty years old. Here is an opportunity to convert their standard induction units into displacement induction units. The entire infrastructure can remain the same, including the air risers, air-handling units, chilled and hot water piping, thermostat and controls, and so on. The induction units were normally placed under the window and have a grille in front for air intake and on top discharging air upward facing glass. The displacement induction units are similar and look identical; the only difference is in their operation. The air discharge grille is located in front of the unit at low floor level. The front low grille allows for air discharge for displacement ventilation. This presents a great opportunity to reduce energy consumption while renovating and retrofitting health care facilities.

PERIMETER BUILDINGS

Displacement induction units are especially suitable in buildings with large perimeter spaces. Such buildings usually are long and slender, with a central circulation corridor and perimeter offices on either side. These buildings generally use perimeter air conditioning units such as fan coil units under the windows. Ventilation air is generally supplied through a central air-handling unit system with diffusers in the ceilings. DIUs installed under the window can produce displacement ventilation in the perimeter spaces, which are generally offices. This can be done for both new buildings and renovations of older perimeter buildings. Older buildings have fan coils and packaged terminal units generally under the window sill, which can be replaced with DIUs. Displacement induction units in these buildings will lower energy use and improve the indoor air quality. The ventilation systems can be coupled with the induction units, eliminating the need for additional ceiling diffusers and ducts. This system also eliminates the need for electrical work, as the units do not have any electrical connections.

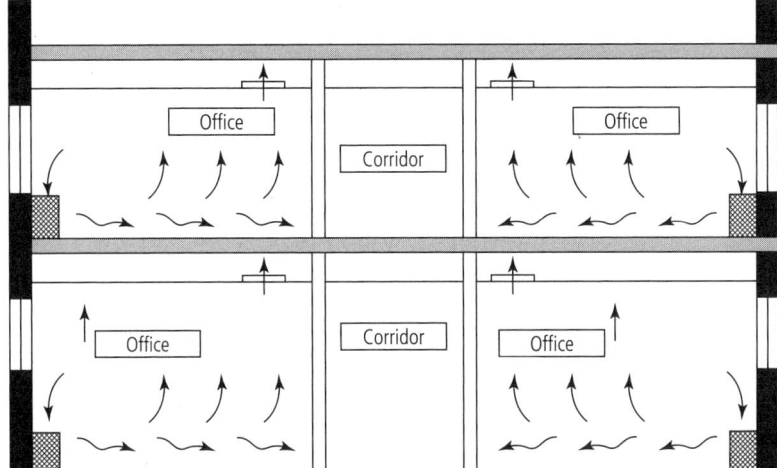

Figure 6–11 Perimeter building displacement induction units. *Asif Syed*

OPERABLE WINDOWS IN BUILDINGS

Displacement induction units are especially suitable for buildings with operable windows, due to the possibility of condensation, which can be collected and drained in DIUs. The fear of using chilled beams in buildings with operable windows is that the occupants may leave the windows open during a humid season, thereby leading to condensation. Though chilled beams have emergency condensate drain pans in some models, it is difficult to drain the condensation with pitched pipes. Therefore chilled beams are generally installed only in buildings with fixed windows, due to the risk of condensation when a window is left open in high-humidity weather. Displacement units are at the floor level and have a condensate drain pan, which eliminates the risk of condensation. The window is generally interlocked with the displacement induction unit's control valve, via a window sash sensor. This permits the closing of the valve when the window is open, saving energy.

CHAPTER 7

High-Performance Envelope

THE BUILDING ENVELOPE IS THE BARRIER THAT SEPARATES THE INDOOR SPACE from the outdoors. The envelope keeps the outdoor elements such as rain, wind, heat, cold, light, and noise from entering the indoors. In most climate zones and in most seasonal climates, the occupant comfort environment usually differs from the outdoors, and the envelope maintains the indoor climate. The conventional approach in most commercial buildings is to provide a solid barrier to keep the outdoor elements out. The building envelope is made up of below-grade foundation walls and basement slab on grade, exterior walls, fenestration (glass and windows), roofs, and skylights. Design of the building envelope is a complex balance of several variables with different expertise such as architectural aesthetics, structural forces resisting wind and building dead and live loads, heat and light transfer for indoor occupant comfort, safety and security, acoustics and sound attenuation, fire resistance, and constructability costs. The building envelope has become a specialized science and art and specialized consultants have emerged who work with architects and structural and mechanical engineers. This science and art calls for an integrated design approach, where input from several different experts and expertise has to come together to an optimum solution. The discussion in this chapter will focus on thermal and light performance, and set aside the safety aspects of the envelope. The performance refers only to the thermal behavior of the façade, but it is important to note that the mechanical engineers by themselves cannot achieve the desired level without participation from architects and structural engineers.

In addition to light and thermal performance of the wall, the floor slabs in buildings with balconies can affect the overall performance of the envelope. Balconies are common in residential buildings; they enhance the quality of living, and also indirectly reduce energy consumption by providing an opportunity to open windows and extend the living to outdoors. However, in extreme climatic conditions they can form thermal bridges.

High performance does not indicate the level of performance, unless it is measurable with a benchmark as base. For the design team and the larger building stakeholders to design a high-performance envelope, the basics of the subject have to be understood. These include factors that affect the selection of envelope, the construction codes and standards, climate, and use of the building. The goal of this chapter is to familiarize the readers with all the issues and factors related to performance of the envelope. Once the basic science of the envelope is understood, it is easy for the readers to get involved in making the envelope better by improving its performance. The technologies identified here are indicative of different approaches. There is no one universal solution that fits all. Each building is unique and requires a thoroughly individualized approach to arrive at the solution. Understanding the science of building envelopes, the code compliance paths, and climates is the best tool that can be provided to the designers to arrive at an envelope that performs better than mandates of construction codes and is measurable.

ENGAGING AND NONENGAGING ENVELOPES

Generally, from the thermal performance point of view, the envelope can be classified into two categories, thermally engaging and thermally nonengaging, based on its interaction with the outdoor environment. An envelope that does not change with the outdoor environment is a nonengaging envelope. An example of a nonengaging envelope is a solid barrier, such as an opaque wall or fenestration. In a nonengaging envelope the occupant or an automatic or passive device has no control to react to the changing needs of the indoor environment due to changes in indoor and or outdoor environments. An engaging envelope could be as simple as an operable window or a shade or blind on fenestration, which allows an occupant or an automatic control mechanism to react to the outdoor environmental elements by engaging with them, to his or her benefit. An engaging envelope is generally more efficient and provides higher performance. In most climate zones, there are seasons when outdoor environments are not far from the desired indoor comfort environment. An effectively engaging envelope allows the occupant to make use of the outdoor environment, reducing the mechanical energy required for heating and cooling. Most commercial buildings generate heat indoors from lights, occupants, and equipment such as computers and server rooms. This heat generation happens irrespective of the outdoor environment of cold winter or hot and humid summer. So both extreme and milder seasons provide an opportunity to engage the outdoor environment by rejecting or capturing heat to and from the outdoors.

HIGH-PERFORMANCE ENVELOPE DEFINITION

A high-performance envelope helps to reduce energy use in buildings. The components that make up the envelope are walls, glazing, mullions, roof, skylights, doors, windows, and slab edge insulation. These components contribute significantly to the energy consumed in buildings for heating and cooling. There is no universal definition for a high-performance envelope; at best, it can be defined as "better than the conventional industry practice" or "better than code mandate." The conventional industry practice is to follow the energy code; most states have adopted some version of an energy code.[1] A high-performance envelope can now be safely defined as one that performs better than the mandatory energy code requires, while consuming less energy than the mandated energy code allows. The envelope performance can be significantly improved by expensive methods, which can save energy but have a long payback and are not practical for most buildings, their owners, or their occupants. The focus of this chapter is on technologies that are energy efficient, practical, and easy to build, and have a reasonable payback or return on investment. To improve performance to "better than code," this chapter also explains what the current codes mandate; knowing this is essential to meeting and exceeding code requirements.

The building envelope uses a significant portion of the energy consumed by the building. The envelope cooling load of a multifamily residential building can be as high as 50 percent of the entire load. Glass has become an important material in construction, as it is suitable in almost any climate. Glass is durable to the extremities or cycles of temperature. Construction cost of glass envelopes is reasonable, making it a highly preferred and constructible material. Above all, it brings in visible light while keeping the elements out. Glass keeps water out of the building more effectively than any other construction material, and this property has led to the use of glass in opaque form where light transmission is not required. Many iconic and signature architectural projects have chosen extensive use of glass in their buildings. With more use of glass, the envelope becomes an important element in the energy performance of the building and it is important for architects and engineers to work together toward a high-performance envelope.

For an office building of approximately 300,000 square feet in zone 4A (mixed climate zone 4 and moist definition—A), the percentage of cooling loads from the building envelope that complies with the prescriptive requirements of the energy code is:

Solar heat gain from glass windows: 14.7%

Glass heat transmission: 2.1%

Wall heat transmission: 1.6%

---

[1] Building Energy Codes Program, U.S. Department of Energy.

Figure 7–1 Percentage of façade cooling load. *Asif Syed*

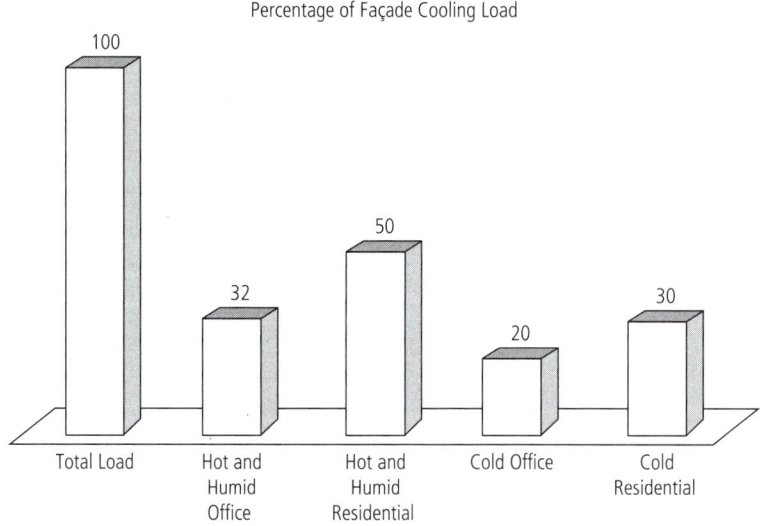

As statistics have shown, glass is a significant contributor to energy consumption in buildings. About 25 to 35 percent of energy consumed by buildings is due to the use of glass. About 10 percent of the total carbon emissions in the United States can be attributed to glass.[2]

The codes that are most commonly used by most states are the International Energy Conservation Code (IECC) and or the American National Standards Institute/American Society of Heating, Refrigerating and Air-Conditioning Engineers (ANSI/ASHRAE) Standard 90.1. The IECC and ANSI/ASHRAE 90.1 are similar in concept and approach, but are not identical. Some states allow compliance with either IECC or ANSI/ASHRAE 90.1. In this chapter, we will use ANSI/ASHRAE 90.1 for the rest of the discussion. There are residential and commercial versions of the code. The residential version of the energy code is for residential buildings that are less than three (3) stories. High-rise multifamily residential buildings fall under the commercial version of the energy code. All nonresidential buildings also fall under the commercial version of the energy code.

MOST COMMON ENERGY CODES: ANSI AND ASHRAE 90.1

The most common energy code followed for building envelope is American Society of Heating, Refrigeration and Air-Conditioning Engineers (ASHRAE) Standard 90.1. The code is also known as ANSI (American National Standards Institute)/ASHRAE 90.1.

---

[2] Bruce Lang, "Energy-Saving Alternatives: ENERGY STAR's New Glass Performance Standards Recognize Superior Energy-Saving Alternatives to Generic Low-e Glass," *Environmental Design & Construction*, May 1, 2008.

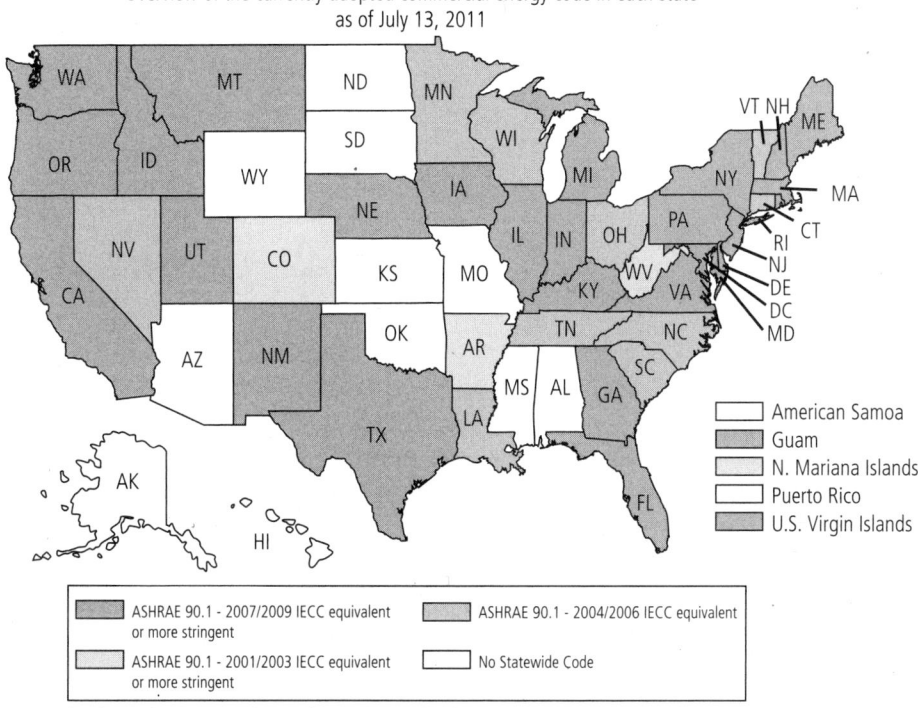

Figure 7–2 Energy code adoption in the United States. *Asif Syed*

ANSI/ASHRAE 90.1 was first developed in 1975 and went through revisions in 1980, 1989, and 1999. Due to the extraordinary interest in energy, the standard is under regular review, with a three-year publication cycle consistent with building codes. The most commonly adopted current standard is 90.1–2007. The latest version is 90.1–2010. In most states the standard is the 2004 version of the code. The 2007 version has raised the bar. The mandated envelope requirements are more stringent, limiting the envelope to 40 percent fenestration. This helps improve the operating efficiency of heating and air conditioning.

CLIMATE ZONES

Code-mandated building envelope criteria are based on climate zones. Prior building codes used heating and cooling degree days. Heating degree days were more popular and available, because most buildings were heated. The cooling energy was not represented accurately for buildings designed with heating degree days methodology. Climate zones more accurately represent both heating and cooling. The climate is classified into eight climate zones and three definitions. The eight climate zones are:

1 – Very Hot

2 – Hot

3 – Warm

4 – Mixed

5 – Cool

6 – Cold

7 – Very Cold

8 – Sub-Arctic

The definition is a subset of the climate zones and identifies the zone by the moisture in the atmosphere. The three climate definitions are:

A – Moist

B – Dry

C – Marine

Every county in the United States is identified by zone and definition. A few examples are shown in Table 7–1.

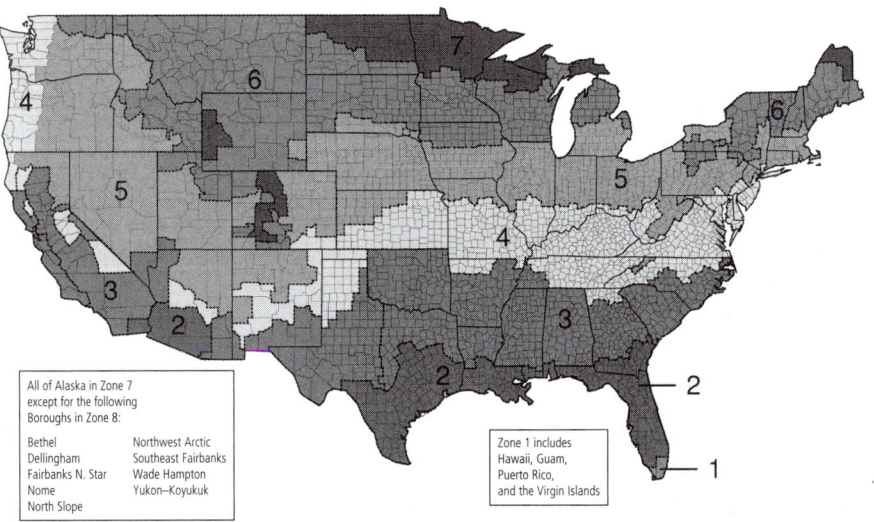

Figure 7–3 Climate zones of the United States.   *Building Energy Codes (resourcecenter.pnl.gov)*

TABLE 7-1 EXAMPLES OF CLIMATE ZONE AND DEFINITION

State	City	County	Zone and Definition
Florida	Miami	Dade	1A
Florida	Orlando	Orange	2A
Georgia	Atlanta	Fulton and DeKalb	3A
District of Columbia	Washington	District of Columbia	4A
New York	New York	New York	4A
Illinois	Chicago	Cook and DuPage	5A
California	Los Angeles	Los Angeles	3B
Washington	Seattle	King	4C
Arizona	Phoenix	Maricopa	3B

COMPLIANCE WITH ENERGY CODES

The most common energy code adopted by most of the United States and by the United States Green Building Council (USGBC) for LEED certification is ANSI/ASHRAE 90.1 for commercial buildings. Under provisions of the American Recovery and Reinvestment Act of 2009,[3] the states had to conform to the International Energy Conservation Code (IECC) for residential construction and to ANSI/ASHRAE 90.1–2007 for commercial buildings, in order to be eligible for federal energy grants. The governors had to certify that their state would adopt an energy code that met or exceeded the provisions of those standards. We will limit our discussion to commercial buildings and discuss compliance with ANSI/ASHRAE 90.1. There are three different ways that can be used to verify compliance with Standard 90.1. See Table 7–2.

The prescriptive path identifies the envelope performance for wall and glazing. The wall and glazing U-values are based on climate zones. The prescriptive path limits the vertical glazing (windows) to 40 percent for all climate zones and the horizontal glazing (skylights) to 5 percent. The percentage of glazing is also commonly referred to as the wall to window ratio (WWR). The wall and glass performance values are tabulated in the code. There are tables for each climate zone and definition. The

---

[3] "States Adopting, Administering ICC 2009 IECC to Receive Energy Assistance Grants," IHS, February 12, 2009, www.ihs.com/news/energy-efficiency/2009/icc-energy-iecc-21209.htm.

TABLE 7-2 OPTIONS TO COMPLY WITH ENERGY CODE

1	Prescriptive path
2	Trade-off option
3	Simulation by the energy cost budget method

tables list the wall U-value of various types of construction. The tables also provide the glass U-value (overall window plus frame) and solar heat gain coefficient (SHGC) value. If the building envelope complies with these values, there is nothing further to be done. However, if any of the prescribed values can't be met, then the next compliance option must be investigated.

The trade-off option permits the trade-off of envelope components, providing more flexibility to the designer than the prescriptive path. Performances (U and SHGC) of envelope components—such as glazing, wall, roof, slab on grade, skylight, and doors—can be traded, as long as the overall effect is the same as or better than that of the prescriptive option. The envelope performance factor (EPF) is used to compare the actual design with the prescriptive path. The EPF has to be less than or equal to the prescriptive path. EPF calculations are defined in the code and can be performed. The calculation methodology is explained in the energy code. The EPF takes into account the impact of the envelope on the HVAC and lighting. There is free software available to perform the trade-off calculations.

COMCHECK AND RESCHECK

The Department of Energy has developed the REScheck (residential applications) and COMcheck (commercial applications) programs to assist the design community with energy code compliance. Both programs save time and effort in the process of documenting compliance. The input includes envelope, lighting, and mechanical systems. The output is a certification statement that indicates compliance or noncompliance. The code officials who receive compliance output, signed and sealed by architects and engineers, can review it with ease. Most building departments accept the certified outputs as part of the building permitting process. Both REScheck and COMcheck are free and can be downloaded from the Web.

The ANSI/ASHRAE 90.1 mandatory requirements for glazing performance for different climate zones are listed in Table 7-3. A higher-performance envelope has to meet and exceed the mandatory requirements. After the discussion of the glazing characteristics, various technologies and architectural features that can exceed the mandatory requirements will be discussed.

TABLE 7-3 ANSI/ASHRAE 90.1 MANDATORY REQUIREMENTS FOR GLAZING PERFORMANCE FOR DIFFERENT CLIMATE ZONES

#	Climate Zone	SHGC	Max % of Vertical Glazing	U-Assembly Curtain Wall (Glass + Frame)
1	1A, 1B	0.25	40%	1.2
2	2A, 2B	0.25	40%	0.7
3	3A, 3B, 3C	0.25	40%	0.6
4	4A, 4B, 4C	0.40	40%	0.5
5	5A, 5B, 5C 6A, 6B	0.40	40%	0.4
6	7, 8	0.45	40%	0.4

SIMULATION BY THE ENERGY COST BUDGET METHOD

Simulation by the energy cost budget method permits evaluation of the building in its entirety or the whole building approach. This includes the mechanical and electrical systems and the entire envelope. This method gives credit for innovations in energy-saving features of one system, which can compensate for energy consumption by another system. The interactive synergies of the mechanical, electrical, and architectural components are accounted for in this method. The annual energy cost of the proposed design has to be less than the cost of the base prescriptive building design. Advanced technologies such as chilled beams, displacement ventilation, radiant cooling, high-performance envelope, and underfloor air distribution, which use far less energy than conventional systems (variable air volume systems per code prescription), can be traded off with other architectural elements.

GLAZING CHARACTERISTICS

There are four important glazing characteristics, two of which are regulated and mandated by code. In addition, the code also mandates the percent of glass in the wall or window wall ratio. The energy code directly mandates the performance of the framing or mullions of the glass. The U-value mandated is the assembly of glass and frame.

To improve the performance of the envelope, the minimum code-mandated values specific to the site and buildings have to be established. These values can vary based on the climate zone and building use. One easy way to improve the

TABLE 7–4 FOUR IMPORTANT GLAZING CHARACTERISTICS

Symbol	Glass Characteristic	Code Mandate
U – glass (center of glass)	Heat transmission coefficient BTU/hr sq. ft. F (glazing only)	No
U – glass + frame (assembly)	Heat transmission coefficient BTU/hr sq. ft. F (assembly)	Yes
SHGC	Solar heat gain coefficient	Yes
VLT	Visible light transmittance	No
LSG	Light to solar gain ratio	No

performance of the envelope is to select a glass with lower U-values and SHGC or have lower window wall ratio. Other ways of improving glass performance is to engage the envelope with the exterior elements. Each of the four glass characteristics are discussed here to understand the variable and their mutual synergies to make the optimum selection of glass.

U–VALUE: HEAT TRANSMISSION COEFFICIENT BTU/HR SQ. FT. F

U-value is a measure of heat transmission from one side of the glass to the other. U-value is the inverse of R-value (U = 1/R). R is the measure of thermal resistance to heat transmission, normally used for insulation. The lower the U-value, the better the performance or lower heat transfer rate. Most commercial building glass is an assembly of multiple glass panes and a space between them and is commonly referred to as IGU (insulated glass unit). Thermal performance or insulating value (U) depends on the thickness of glass, number of panes in the assembly, spacing between the panes, and type of gas or vacuum between the panes. As a rule of thumb Table 7–5 lists the U- and R-values of glass. As the U-value decreases the cost of the IGU increases.

It is important that selection of glass decisions be made not just by glass U-values, but the combined assembly of glass and frame U-values. The prescriptive mandate for thermal conductance is for both glass and frame U- (assembly) value. The U- (assembly) values can be obtained from the energy code, specific to the site.

Glazing U-Value and Glazing Assembly U-Value

Glazing U-value is the centerline U-value of glass itself, and does not include any heat loss from the frame that holds the glass in place in the wall assembly. The glazing U-value mandated in the code is the assembly value. Generally, the overall assembly U-value is lower than the glazing-alone U-value. For example, for double-glazed glass in a standard aluminum frame, the overall U-value is twice the glazing U-value.

TABLE 7-5 IGU U-VALUES

No.	IGU Assembly	U	R
2	Single pane glass	1.1	0.9
2	Clear glass with air gap	0.5	2
3	Clear glass with argon filled in gap	0.33	3
4	Low E on # 2 surface	0.25	4
5	Triple glazed with Low E on #2 and #4 surface	0.2	5
6	Vacuum or multichambered glass	0.08	12

U glass = 0.25 Btu/hour sq. ft. F

U overall = 0.5 Btu/hour sq. ft. F

In order to reduce the overall U, it is important to reduce the heat transfer in the frame or the mullion of the glass. Some of the methods to reduce the frame or mullion U-values are:

1. A thermally broken frame or mullion

    A thermal bridge or direct metal connection from outdoors to inside is the primary cause for low thermal performance of the window or curtain frame. A thermal break is achieved by splitting the frame into two separate elements for outside and inside, joining them to glass with an insulating material. The insulating material commonly used is polyurethane, which has much lower conductivity than aluminum. Another problem with thermal bridges is condensation. In cold climates, sections of exterior wall can get cold and form condensation and frost, especially in buildings that are humidified in winter. Condensation leads to other indoor air quality problems of mold and bacteria growth. Avoiding a thermal bridge also has indoor air quality benefit.

2. Nonmetallic spacers

    In most cases, windows are double pane insulating glass units (IGU). The two layers of glass are generally separated by a metallic element. Aluminum was common but it has been replaced with steel. Steel's thermal conductivity is about four times lower than aluminum. However, a nonmetallic spacer's thermal conductivity can be about one hundred times less than steel, giving the best performance.

SOLAR HEAT GAIN COEFFICIENT (SHGC)

Solar heat gain coefficient is the fraction of heat from the sun that enters the envelope through glazing. It is a number from 0 to 1. The lower the SHGC, the less energy enters the building. Higher SHGC indicates that more energy enters the building. An SHGC of 0.4 indicates that only 40 percent of the solar energy incident on the window enters the glazing, and 60 percent is reflected or dissipated outdoors. Clear glass has an SHGC close to 1 and opaque glass 0. The energy code mandates the SHGC coefficient for all climate zones, the sample of the prescriptive values are shown in Table 7–3. The SHGC of glass is reduced by adding color tints, reflective coatings, spectrally selective low E coatings. The coating is applied between the glazings of double-glazed windows. The most effective way to reduce the SHGC is with spectrally selective low E coatings. The spectrally selective low E coating, while reducing the solar energy transmitted through glass, maximizes the beneficial visible light, as shown in Figure 7–4. The SHGC, while reducing the harmful heat gain in summer, may reduce the beneficial heat gain in winter. The process of selecting the heat gain coefficient has to achieve a balance between summer benefits and winter losses. Different types of low E coatings are available such as low solar gain, moderate solar gain, and high solar gain. The high solar gain coatings work well in cold climates and low solar gain coatings work well in hot climates.

A rule of thumb range for SHGC for various types of glazing is shown in Table 7–6. The values are only indicative to demonstrate what makes the IGU lower the SHGC.

VISIBLE LIGHT TRANSMITTANCE (VLT)

Visible light transmittance is a measure of the amount of visible light that passes through the glazing. Visible light transmittance (VLT) or visible transmittance (VT) expressed as a percentage is the fraction of the visible spectrum of light that is transmitted through the glazing. A higher VLT means that more light enters the building.

TABLE 7–6 IGU SHGC VALUES

No.	Glass Assembly	SHGC
1	Single pane uncoated	0.8
2	Double pane uncoated	0.7
3	Double pane tint low E (moderate)	0.4
4	Double pane tint low E (low)	0.27
5	Double pane mirror low E (low)	0.22

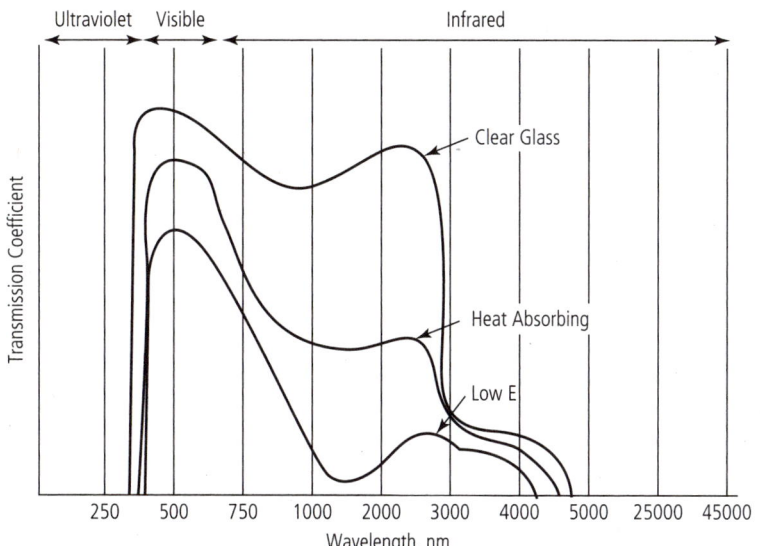

Figure 7–4 Solar spectrum and low E. *Asif Syed*

VLT is important to a high-performance envelope, as it has an impact on daylight harvesting. Higher VLT is desirable for daylight harvesting, which reduces the energy used for lighting.

LIGHT TO SOLAR GAIN RATIO (LSG)

Light to solar gain ratio (LSG) is the VT value divided by the SHGC value. Higher LSG is indicative of lower solar heat gain and higher visible light transmittance. An LSG greater than 1 indicates that more light is transmitted than heat. A low LSG number indicates that very little light enters the space, defeating the purpose of having windows for a view of the outdoors and daylighting. The LSG has become an important measure of building envelope performance. Higher LSG is possible with low E coatings and spectrally selective and tinted glasses.

Table 7–7 shows the effects of various combinations of coatings on clear, tinted, reflective, and spectrally selective glazing. Introduction of low E coating reduces the U-value. Reflective glass reduces VLT drastically while reducing SHGC. Introducing low E on color tint increases LSC from 0.92 to 1.40 (52%). Spectrally selective and low E give the maximum performance with low U (0.28), reasonable VLT (53%), low SHGC (0.3), and high LSG (1.77). The values in the table are a general rule of thumb and demonstrate the effect of the variables. Specific project variables will be different and there is no universal solution that fits all projects. An integrated collaborative approach is required to select the right glass and participation from architect, engineer, lighting designer, and façade consultant is required.

TABLE 7-7 EFFECTS OF TINT, LOW E COATING ON U, VLT, SHGC, AND LSG

No.	Glass Type – Double Pane	Glass U	VLT—%	SHGC	LSG
1	Clear	0.49	79	0.70	1.13
2	Reflective	0.48	23	0.21	1.10
3	Reflective and low E	0.27	25	0.20	1.25
4	Color tint (green)	0.49	47	0.51	0.92
5	Clear low E	0.28	70	0.45	1.56
6	Color tint and low E	0.28	45	0.32	1.40

HOW TO EXCEED THE MANDATORY CODE PERFORMANCE

As seen from the above discussion in this chapter, exceeding the code mandated performance to achieve a high-performance envelope is a combination of several variables. All aspects of all variables have to be evaluated specific to the building geometry, site, occupancy, building use, and so forth. Therefore, understanding the basic definition of the variables and their interaction with each other is essential to design a façade that exceeds the code mandate performance. It is an integrated approach involving architect, engineer, façade consultant, and user. Technologies and architectural features of the envelope discussed further in this chapter generally benefit and allow a building design to exceed the required minimum code performance and create a high-performance envelope. Adopting these technologies without comprehensive energy evaluation of the entire building with site-specific conditions is not recommended. Only an energy analysis can establish the true benefits of individual elements. These elements are:

1. Operable windows
2. External shades, overhangs, and fins
3. Solar-responsive internal blinds or shades
4. Double-skin envelope
5. Triple-pane glass
6. Thermal mass
7. Green Roofs

OPERABLE WINDOWS

Operable windows are a simple approach to reduce energy, exceed energy code compliance, and also provide the occupants with the ability to control their environment. Operable windows are desired by most building occupants, as this feature allows them to engage with the outdoor environment and control their indoor environment. Most engineers and building operators, on the other hand, see operable windows as features that waste energy. It is true that when operable windows expose indoor HVAC systems to the outdoor environment, the amount of cooling or heating the HVAC systems have to do is greater than normal, causing them to consume more energy. Both building occupants and engineers are correct in their respective desires to have operable windows and have sealed façades or envelopes. The problem of excessive energy consumption can be converted into a feature of energy savings by shutting off the HVAC systems when the windows are open. Now engineers have to figure out a system that automatically detects an open window and shuts off the HVAC system. Automatic shutdown of HVAC systems is not a common practice nor is a conventional system with standard products available. With the abundant automatic controls of HVAC systems via the building management and security (BMS) systems, the shutdown of individual zones of the HVAC system can be easily achieved. The shut-off is easily automated with a window sensor, which can be a simple device similar to a window security alarm sensor. The window sensor detects if the window is open or closed and sends feedback to the building HVAC control system via the

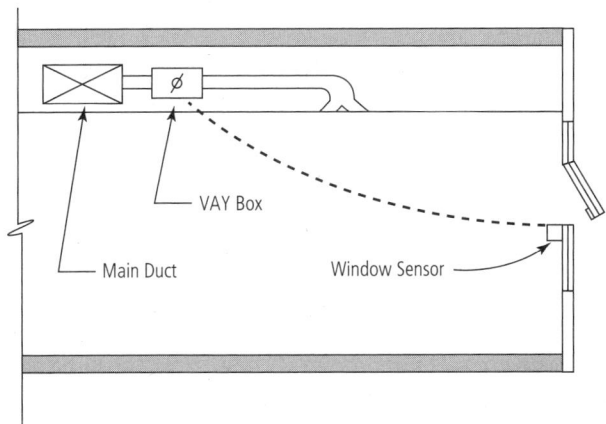

Figure 7–5 Operable window sensors. *Asif Syed*

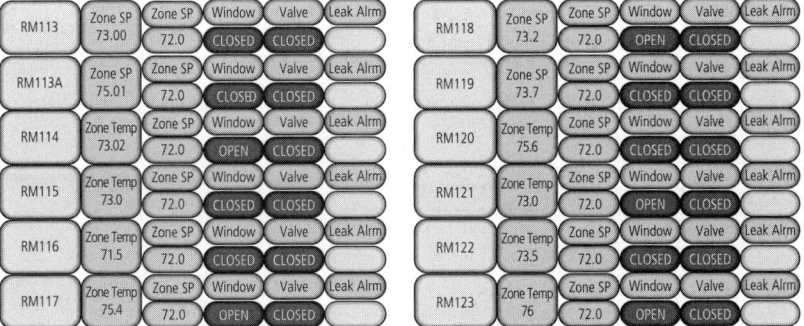

Figure 7–6 Operable window controls. *Asif Syed*

BMS system. With this feature, when a building occupant opens a window, the HVAC system automatically detects this condition and shuts off the HVAC system associated with the window. Educating, involving, and familiarizing the building occupants with these controls will make them actively participate in the process, which helps reduce energy use.[4] In almost all low-rise buildings, operable windows integrated with the HVAC systems can provide benefits. However, in high-rise construction, the stack effect must be carefully studied. The stack effect is similar to the chimney effect, whereby the hot air rises up the building (in summer) and cold air comes down the building (in winter). The stack effect is very strong in high-rise buildings and can make air enter the building or leave the building in an uncontrollable manner. Proper isolation of the perimeter or exterior spaces with interior elevator hoistways and stairs can minimize the stack effect air flow into and out of the building. The stack effect is based on the difference between indoor and outdoor environment. In seasons when the outdoor temperature is close to indoor environment, the stack effect is minimal.

EXTERNAL SHADES AND OVERHANGS

External shades and overhangs are a simple approach to reduce energy consumption in the building. External shading devices are not part of the mandatory requirements of the energy code. Therefore, the benefits obtained from external shading devices will go toward making the envelope better than the code mandate, making them high-performance elements of the façade. There are two major ways that external shades benefit performance or reduce energy consumption:

- Reduction of direct solar gain
- Reduction of light glare

The reduction in energy use by reduction of solar gain is straightforward. Any energy that does not enter the building does not have to be taken out of the building via cooling through HVAC systems. This reduces energy consumption and reduces the peak demand for electricity to run chillers, pumps, cooling towers, fans, or any other components of the HVAC system.

The reduction in energy use from reduction in glare is indirect. Excessive glare in commercial buildings interferes with normal operations such as working with computer monitors and projectors in conference rooms. In order to reduce glare, blinds or blackout shades are rolled down, leading to a reduction in daylight and natural views. The reduction in daylight leads to an increase in energy use from turning on the building lights. The external shades and overhangs, however, reduce glare, reduce energy consumption, and increase daylight views.

---

[4] Allan Daly, "Operable Windows and HVAC System," *HPAC Engineering*, December 1, 2002.

The external shades and overhangs must always be evaluated along with, or in combination with, the performance of the glazing used. These two elements of the façade act as one integrated element. The maximum effect of external shades and overhangs is achieved when they are used with clear glass. However, the energy benefits of external shades and overhangs are not completely eliminated with high-performance low E glass. The reduction in energy is tabulated in Table 7–8.[5]

TABLE 7–8 REDUCTION IN ENERGY AND GLARE WITH OVERHANGS

No.	City/Location	3.3 Feet of Overhang with Glass Type	Glass SHGC	Glass T-vis	Energy Savings %	Glare Reduction %
1	Minneapolis	Clear glass	0.6	0.63	15.9	8.21
		Low E tint	0.39	0.36	11.8	9.2
		Low E clear	0.22	0.37	8.7	9.2
2	Washington, D.C.	Clear glass	0.6	0.63	15.4	8.2
		Low E tint	0.39	0.36	11.9	9.3
		Low E clear	0.22	0.37	8.2	9.2
3	Chicago	Clear glass	0.6	0.63	17.0	8.0
		Low E tint	0.39	0.36	13.6	9.0
		Low E clear	0.22	0.37	8.8	9.0
4	Houston	Clear glass	0.6	0.63	19.0	8.4
		Low E tint	0.39	0.36	14.0	9.5
		Low E clear	0.22	0.37	8.9	9.5
5	Phoenix	Clear glass	0.6	0.63	21.0	7.1
		Low E tint	0.39	0.36	17.7	7.8
		Low E clear	0.22	0.37	11	7.8
6	Los Angeles	Clear glass	0.6	0.63	21	7.4
		Low E tint	0.39	0.36	17.3	8.1
		Low E clear	0.22	0.37	12.1	8.1

[5] John Carmody and Kerry Haglund, *External Shading Devices in Commercial Buildings: The Impact on Energy Use, Peak Demand and Glare Control,* Center for Sustainable Building Research, University of Minnesota/AMCA International, 2006.

Figure 7–7 External shading at Petronas Twin Towers, Kuala Lumpur, Malaysia. *Asif Syed*

Figure 7–8 External shades at Newseum, Washington, D.C. *Asif Syed*

SOLAR-RESPONSIVE BLINDS AND SHADES

There are blinds and shades in almost every building. Reducing direct sunlight and preventing excessive diffused sunlight from entering the buildings are important to almost any building. The shades are either manual or motorized; motorized shades are becoming common in commercial buildings. In spite of motorizing the blinds, their control in most buildings is by the occupant via a wall-mounted switch. Unless the occupant consciously closes the blind every time there is direct sun on the window, the benefit from motorized blinds is limited. To maximize the use of motorized blinds, they can be automated to respond to sun and light. This automating technology is extremely cost effective, as the major cost of the motorization of blinds and electrification costs are already there. The only additional cost is automation. Developing a control algorithm is a challenge, as it is not a common practice in the industry. Such an algorithm has to be developed by engineers.

Independent of the shading control, most buildings—especially those with sustainability goals—have a daylight-harvesting system, with photo sensors and dimmers. These systems are normally independent of the blind position. This system dims the lights to maintain predetermined light levels. If the two systems are not integrated they will likely fight each other. The shades may close to reduce the solar heat gain and the light levels will increase to compensate for the closing of the blinds. The benefits of both the systems can be maximized by integration.

The daylight-harvesting system and the solar radiant system can be integrated into one system with simultaneous:

1. Motorized shading, which reduces solar heat gain
2. Daylight-harvesting system, which allows optimum light to reduce solar energy.

The algorithm to integrate the two systems is complex and needs to be customized. Developing such an algorithm presents a challenge as the façade environmental conditions constantly vary. The solar heat gain and visual light transmittance varies hourly based on the position of the sun and seasonality. The light levels can be easily measured with a photo sensor, but the thermal performance has to be calculated based on outdoor conditions such as temperature and solar radiation intensity. Cloud cover and light diffusivity has to be factored as well. With all the input the system has to establish a position of blinds that will:

1. Optimize heat gain (allowing solar heat gain in heating mode, and reducing solar heat gain in cooling mode)
2. Control light glare
3. Optimize daylight

The input and outputs of the control system are shown in Table 7–9.

TABLE 7-9 INPUTS AND OUTPUTS OF SOLAR-RESPONSIVE SHADING SYSTEM

Input	Output
Outdoor temperature	Shade positions
Indoor temperature	
Indoor and outdoor photo sensor light levels	
Sun position (time – day – month)	
Window direction (site geometry)	
Façade thermal performance	
Façade VLT	
Solar radiation intensity	

Based on the outdoor and indoor temperature, the algorithm will establish if heating or cooling is required at the perimeter zone. When heating is required, the solar shades will open to allow heat gain into the space at a preset glare level, and dim the lights via the dimmers. When cooling is required at the perimeter zone, the shades will close to reduce the solar heat gain, maintaining the preset light levels to reduce the use of electricity for lights. In both conditions, an optimum balance has to be established between light energy consumed and heat loss or gain.

DOUBLE-SKIN ENVELOPE

The fundamental principle of all double-wall façades is based on an intermediate cavity between an inner and outer façade. The goals of the cavity are to reduce the impact of the external environment and to take advantage of the external environment. The intermediate cavity enhances the building energy performance by:

1. Harvesting energy from the exterior environment
2. Conserving the energy of the indoor environment

The non-energy-related benefits of the double-wall façade include protection from the elements and acoustic buffering. The discussion in this chapter is limited to energy-related benefits.

The cost of double-wall façades is an important factor in the decision-making process. It is obvious that most double-wall façades are expensive and also take away more of the building's buildable footprint. Irrespective of setting, local zoning ordinances, or regulations governing the floor area ratio (FAR) or floor space index (FSI), an 18-inch to 30-inch space along the perimeter will be lost. The capital cost of construction and the ongoing cost of real estate have to be factored and compared with the energy savings from higher performance that reduce the utility costs. The payback, as documented by David Stribling and Byron Stigge,[6] can vary from twenty to ninety years, depending on the climate and cost of utilities. Most of the investment community and building developers are not ready for such long paybacks; that is why double-wall façades are not used in most buildings.

There are four possible methods of engaging the cavity of the double-wall façade (see Figure 7–9). The façade is normally built to engage in all possible combinations of these four methods. Each arrangement has a benefit at a particular time of year, indoor temperature, outdoor temperature, day and night cycle, and the like. Normally, motorized dampers or manual dampers are installed in the interior and exterior façade at low and high elevations (see Figure 7–10). The motorized dampers

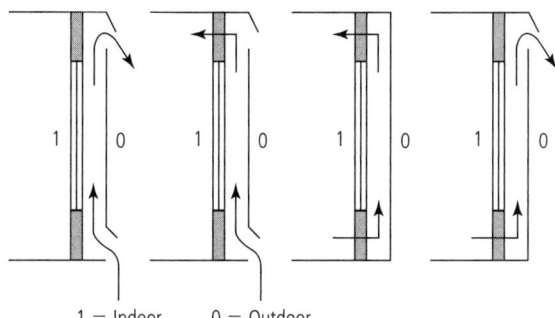

Figure 7–9 Double-wall façade air flows. *Asif Syed*

1 = Indoor   0 = Outdoor

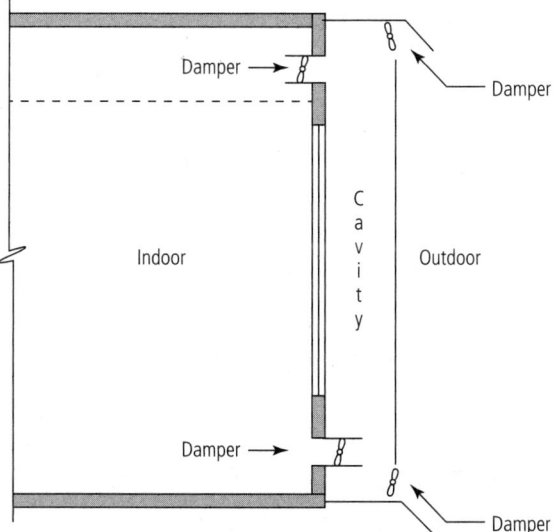

Figure 7–10 Double-wall façade dampers. *Asif Syed*

---

[6]David Stribling and Byron Stigge, "A critical review of the energy savings and cost payback issues of double facades," Chartered Institution of Building Services Engineers (CIBSE)/ASHRAE Conference, 2003.

are normally connected to the building management system, which automatically opens and closes them based on the thermal performance desired at a particular condition. The four methods are:

1. Outdoor air buffer
2. Outdoor air supply
3. Indoor air buffer
4. Indoor air exhaust

To measure the performance level with the baseline to establish high performance and energy savings, a 24/7 and 365-day energy simulation has to be performed. The double-wall cavity presents an additional challenge as most energy simulation software does not have algorithms for the four methods listed here. To establish a realistic performance level of the double-wall façade, engineers have to develop these complex algorithms. The algorithm variables include:

1. Combined thermal performance of both façades
2. Combined light transmission performance of both façades
3. Buoyancy forces in cavity and their impact on thermal performance
4. Air transfer into and out of the cavity from indoors and outdoors
5. Air flow balance between air entering and leaving the cavity with the building ventilation fans

OUTDOOR AIR BUFFER

The intermediate cavity acts as a greenhouse and captures heat. The placement of heat-reflective shades, louvers, blinds, glass film, and the like, can be used to maximize the capture of heat or to increase the greenhouse effect. Ventilation of the cavity is provided by thermal and buoyancy effects within the cavity. The circulation of air within the cavity increases the convective heat transfer, reducing the temperature in the cavity and reducing the thermal conduction between the cavity and the indoors. The outdoor air buffer does not introduce outside air into the building, because only the external façade has an opening to circulate air with thermal and buoyancy effects.

OUTDOOR AIR SUPPLY

An outdoor air supply can introduce ventilation air directly into the space. The openings into the space can be either operable windows or openings with automatic dampers. Operable windows provide occupant control and require interlock with HVAC systems.

An outdoor air supply can be introduced into the building to take advantage of the nighttime cooling of the building. When the temperatures drop in the night, the outdoor air can be introduced into the building to cool the thermal mass, thus creating a thermal inertia. This can be done with either natural or mechanical ventilation. This application can prove to be of benefit in climates with high diurnal temperature change, such as a desert environment. Normally, this has to be coupled with the thermal mass of the building. The higher the thermal mass of the building elements, such as floors, slabs, and walls, the higher the thermal inertia will be.

Outdoor air supply can also be introduced to take advantage of the heat captured in the intermediate cavity in sunlit conditions. During colder winter months, especially on the south façade, hot air from the cavity can be introduced into the building. This offsets the perimeter heat loss, reducing the heating requirements of the building. For most commercial office buildings, heating energy is limited to only the perimeter spaces. Almost all interior spaces require cooling due to heat generated from equipment, lights, and computers. The interior heat is generally rejected to the outdoors with central air-handling systems. Normally, there is no mechanism to utilize the interior heat to offset the heat loss at the perimeter. There are some technologies, such as water source heat pumps, that can transfer the heat from interiors to the perimeter. But water source heat pumps are not as common as conventional variable air volume (VAV) systems with central air-handling units. Double-wall façades provide an opportunity to reduce the perimeter heat loss. The double-wall façades do not move the heat from interiors to the exterior, but reduce the heating of the perimeter.

Because an outdoor air supply system provides openings into the building, its application has to be carefully evaluated for stack effect in high-rise buildings. Outdoor air quality also has to be carefully evaluated prior to introducing unfiltered outside air into the building.

INDOOR AIR BUFFER

The indoor air buffer engages the double-wall cavity with the indoor space. This arrangement allows the indoor air to circulate between the space and the cavity. The closing of the outside wall of the double wall provides an excellent opportunity to harvest energy from the exterior environment. The cavity's greenhouse effect is much stronger with this arrangement, because the thermal buoyancy effect prevents heat from being lost through convection of air to the outside environment. The thermal buoyancy effect produces natural forces to circulate air in winter. The colder, heavier air enters the bottom of the cavity from the space and rises to the top of the cavity, then gets back into the space. This effect can be reversed at night, when there is no sun and therefore no greenhouse effect. Automatic dampers are almost essential to take advantage of this process.

The indoor air buffer concept works well in high-rise buildings because the building is not porous to the outdoors, which increases the air flow caused by the stack effect; however, at every floor the cavity has to be separated from the floor above and below.

The heat captured in the cavity can be used directly in the HVAC system. In buildings with few floors (low-rise), the outdoor air can be drawn from the cavity, reducing the heating required to heat the air.

INDOOR AIR EXHAUST

The indoor air exhaust also presents a good opportunity to engage the façade to reduce the impact of the outdoor environment. In most HVAC systems there is excess air to be spilled. Normally, this is outside air or fresh air brought into the building for ventilation. This air in most cases is about 20 percent of the total air circulating in the building. This excess air is normally ducted back to the air-handling system and spilled out through a louver. This air can be spilled out at the perimeter of the building through the double-wall façade. This excess spill air, while passing through the double-wall cavity, reduces the impact of the external environment. In winter, the excess spill air is warmer than the outdoor air, producing a buffer between the inside space and the outdoors. In summer, it has the opposite effect.

Indoor air exhaust also offers an opportunity to provide nighttime cooling of the building. In climates such as the desert environment, where the diurnal temperature effect is very high, cooler nighttime air can be used to cool the thermal mass of the building. In environments where outdoor air quality is poor, air cannot be directly

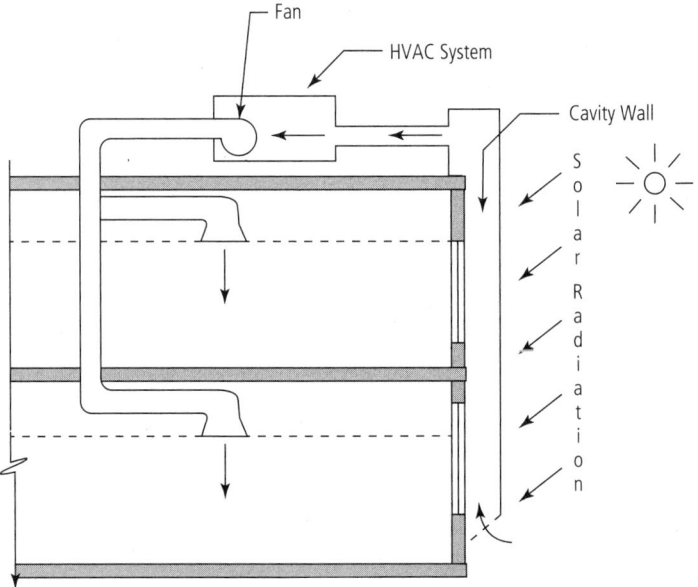

Figure 7–11 Heat harvesting from double-wall façade to HVAC system. *Asif Syed*

introduced into the occupied spaces of the building. Here the outdoor air can be introduced through the air-handling unit (AHU) systems, which have effective filtration. The quantity of air circulated is about 20 to 30 percent of the normal maximum capacity. The lower air flow through a larger duct system significantly reduces the fan energy, making the system more effective.

TRIPLE-PANE WINDOWS

Triple-pane windows have three layers of glass. The three layers of glass provide six surfaces: outside of outer pane (surface 1), inside of outer pane (surface 2), outside of intermediate pane (surface 3), inside of intermediate pane (surface 4), outside of inner pane (surface 5), and inside of inner pane (surface 6). The insulating value of the glass is proportional to the distance between the panes. Panes that are too close together or too far apart do not significantly improve the thermal performance. The optimum distance avoids convection currents between the panes. The distance between the panes can vary from 1 to 1½ inches, depending on the thickness of the panes. Generally the appropriate spacing results in higher thermal performance. However, the three panes make the window or curtain wall system heavy. In a window system with operable windows, the heavier window is a disadvantage, making it more difficult to open and close the window. In a curtain wall system, the three panes make the glass heavy, requiring stronger and bigger frames, which have higher thermal conductance. The thermal conductance of the frame can be reduced by thermally insulating frames or mullions, where insulating materials are used to break the metal continuity from inside to outside. The overall assembly performance of combined glass and frame has to be evaluated to truly establish the benefit of triple-pane glass. Triple-pane windows also cost more than double-pane, so the cost/benefit analysis between energy savings and capital cost has to be evaluated. In climatic zones with extended periods of cold, triple-pane glass generally increases the payback.

The disadvantage of heavy weight created by the third pane can be overcome by a suspended film between the two panes creating the third pane.

Triple-pane glass applications are unique and require careful evaluation weighing climatic zone, cost, thermal performance, aesthetics, and acoustics against one another. In very cold climates, if there is a desire for large windows or clear glass, triple-pane generally pays back the higher cost. One such application is the Helsinki Concert Hall;[7] the cold climate and the desire to have glass architecture were key to the decision to select triple-pane glass. Airport terminal buildings, which require very high sound-insulating qualities, are good applications; here, the sound attenuation and thermal performance can be integrated.

---

[7] "Transparent Music Hall Uses Triple Glass Panes," published January 2011, www.interpane.com/helsinki_music_hall_537.html?sprache=englisch.

THERMAL MASS

Thermal mass reduces instantaneous load by absorbing a portion of the load to be released at a later time. At the later time, the instantaneous load is reduced, thereby offsetting a fractional portion of the building peak cooling load. In addition, in colder climates the building mass can absorb the heat during the daytime hours, store it, and transfer it in the evening or nighttime colder hours. The results of one study[8] indicate that modeling the thermal properties of material with higher mass, such as concrete, can reduce energy consumption, resulting in one (1) energy point for LEED 2.2 Energy and Atmosphere Credit. Normally, energy simulation methods have to be adopted to do such a study. The savings in energy is compared to a baseline system with standard steel construction. Air conditioning peak load in buildings can be reduced by 10 to 15 percent in buildings with a higher mass in walls and floors.

The processes in which energy and environmental benefits are obtained are different, and are explained in the subsequent sections. The brief differences are as follows (see Table 7–10):

REDUCTION IN PEAK COOLING LOAD

Reduction in peak cooling load does not necessarily mean that there is a reduction in energy consumption. Thermal mass reduction in peak load is due to shifting of the load into the nonpeak times adjacent to the peak loads. In this process, the cooling load occurring at the peak hour is absorbed by the thermal mass, such as the exterior walls and floors, and is dissipated later after the peak hour. The process is sometimes intentionally reversed by precooling the concrete mass before the peak cooling load

TABLE 7–10 REDUCTION IN PEAK DEMAND AND REDUCTION IN ENERGY

Reduction in Cooling Load Peak Demand with Thermal Mass	Reduction in Energy with Thermal Mass
Thermal mass does not engage a diurnal cycle.	Thermal mass engages a diurnal cycle.
There is no reduction in energy (same fossil fuel burned).	Reduction in energy occurs (less fossil fuel burned).
Reduces the electrical utility costs by avoiding high peak demand rates.	Reduces the electrical utility costs by avoiding consumption.

---

[8] Medgar L. Marceau and Martha G. VanGeem, "Modeling Energy Performance of Concrete Buildings for LEED-NC Version 2.2: Energy and Atmosphere Credit 1," Portland Cement Association R&D Serial No. 2880a.

occurs and dissipating the load at the peak cooling load hour. Both precooling and dissipation at peak hour and absorption at peak hour and dissipating later only transfer the heat load—they do not reduce it. Usually, peak cooling load occurs in the summer months of July or August at 2:00 to 4:00 PM on western exposure in the north latitude locations. A computer simulation of the building provides peak cooling times for all exposures. Most simulation software allows the designer to vary the thermal mass characteristics of the envelope. Table 7–11 shows the shift in the cooling load for a 150,000-square-foot building located in the suburban Washington, D.C., area.

Reduction in peak loads does not reduce energy consumption, but is not completely without benefits. The benefits include reduction in instantaneous demand for electrical power. Most electrical power grids are at their full capacity, and the peak loads add an additional burden; shifting the loads to off-peak hours reduces the instantaneous demand. Huge costs of upgrading the utility infrastructure to meet the peak demand can be avoided. Also, most power grid systems fire up their worst

TABLE 7–11

Partition Type	Internal Thermal Mass	Cooling Load Reduction
0.75 inches gypsum frame	5 to 6 pounds per square foot	Base case
4-inch-thick lightweight concrete blocks	40 to 50 pounds per square foot	8%

Building type: Office building
Net air conditioned area = 150,000 square feet
Number of floors = 5
Location = Washington, D.C.
Percent of glass to wall area = 40%
Length of partitions used per floor = 500 feet
Height of partition = 10 feet

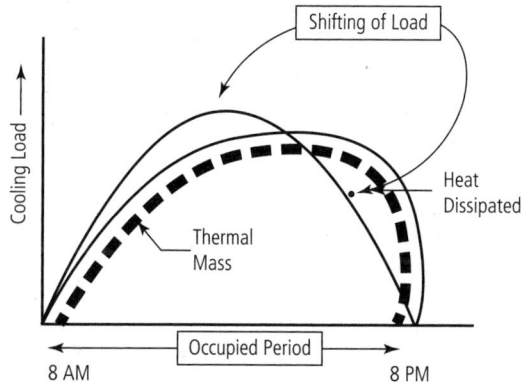

Figure 7–12 Reduction in peak cooling load. *Asif Syed*

or most inefficient or most polluting power generation equipment during the peak hours. Any reduction in peak demand can significantly reduce greenhouse gas emissions. Finally, the cost of electricity is the highest at the peak load hours. Shifting the cooling load to precooling and later heat dissipation reduces the electric utility costs.

REDUCTION IN ENERGY WITH THERMAL MASS

When the thermal mass of the envelope engages with a diurnal cycle, the heat from the daytime solar energy can be stored and released during the night for heating. Similarly, the nighttime coolness can be stored in the envelope mass and released during the day for cooling. The diurnal engagement of thermal mass is possible only when there is a critical mass level that can store sufficient energy at changing conditions—as shown in Figure 7–13—and later release the energy. Thermal mass capacity is defined as heat stored in the material due to the change in temperature, whereas diurnal thermal mass capacity is defined as heat stored and returned in a cycle of 24 hours. A 24-hour simulation or hour-by-hour calculation is required to establish the effectiveness of thermal mass.

TABLE 7–12 THERMAL DENSITY OF MATERIALS

Material	Thermal Density
Concrete	High
Concrete block	Average
Insulation	Low

Figure 7–13 Thermal mass and diurnal temperature. *Asif Syed*

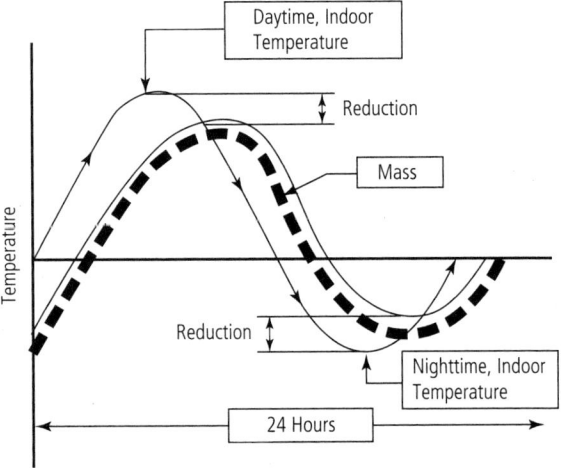

Thermal mass properties are critical to the benefit of thermal inertia, or the flywheel effect. High density and optimum thermal conductivity are required. Concrete is a good example of good thermal mass. High thermal conductivity is a disadvantage, as it conducts heat too fast. Low conductivity is also a disadvantage, as it takes too much time to store heat.

Location of thermal mass must allow the flow of heat by conductivity from outdoors to inside the building. Insulation between the thermal mass and the indoors reduces the effect or benefits of thermal mass.

Engaging the thermal mass with the diurnal cycle reduces energy consumption by avoiding the heat or cooling load required. This benefit results from heat energy transferred from night to day. The reduction in cooling load here translates into avoidance of energy consumption or burning of fossil fuel.

CHAPTER 8

Thermal Energy Storage

THERMAL ENERGY STORAGE IN BUILDINGS DATES BACK TO THE EARLIEST human settlements. Houses built by indigenous people all across the world, in hot and moderate climates, demonstrate some form of thermal energy storage. The most common and easiest form of thermal energy storage uses the diurnal temperature change. Diurnal temperature difference is the difference between the minimum at night, usually lower, and the maximum during the day, usually higher. This leads to colder nights and warmer days. The colder nights provide a "night cool," which can be used to cool the hot days, if the energy can be stored and released. The opposite is also possible—storing the warmth of the day to heat the nights. The indigenous homes stored the night cool in the mass of walls, floors, and roofs. This method can be used in modern and present-day buildings. Using this method of energy storage and transfer can significantly reduce the energy consumed by buildings to provide human comfort. About 33 percent of all energy consumed in buildings goes into heating and cooling.[1] About 40 percent of all energy produced in the United States goes into buildings.[2] About 13 percent of all energy produced in the United States goes into space heating and cooling. The cliff settlements of Mesa Verde in Colorado are an example of thermal energy storage.[3] The cliff dwellings were built to harness the solar energy. In winter the sun warmed the walls of the dwellings, the energy was stored in the mass of the walls, and at night transferred to the space. In summer the location

---
[1] U.S. Department of Energy, Buildings Energy Data Book, 2006 U.S. Buildings Energy End-Use Splits.
[2] National Institute of Building Sciences, Whole Building Design Guide, 2009.
[3] Kelly Hart, "Messages from Mesa Verde," greenhomebuilding.com.

of the dwellings protected them from direct sunlight. These basic principles can be used in present-day buildings. In modern-day buildings the thermal energy storage can be from natural or renewable resources or nonrenewable resource energy can be stored. The nonrenewable energy storage does not reduce the fossil fuel or any other sources of energy consumption, it is merely a better management of the energy production and consumption process.

Stored thermal energy can be from either:
1. Renewable energy sources
2. Nonrenewable energy sources

Both of these methods provide an opportunity to reduce energy costs, but only renewable energy storage reduces energy consumption from burning of fossil fuels, which is the most common form of energy. Renewable energy storage is a true form of energy reduction, whereas nonrenewable energy storage only shifts the loading and reduces the demand for energy for peak hours. The benefits of nonrenewable energy storage include lower-cost energy during the night, but inefficient storage may sometimes use more energy than is consumed during the daytime. So the only true benefit of storing nonrenewable energy is a reduction in the cost of energy, not a reduction in the use of energy. Ice or chilled-water storage is the most common form of nonrenewable energy—commonly known as thermal storage. Renewable energy storage methods are less common but are gaining popularity due to their sustainability benefits.

RENEWABLE ENERGY STORAGE

Renewable energy, by definition, is replenished by natural resources such as wind, sunlight, tides, and rain. There is abundant renewable energy around buildings. Some forms of energy are available on a diurnal basis, and others on a seasonal basis. These diurnal and seasonal swings in energy present the greatest opportunity to use this source for the energy needs of buildings. The storage of this energy can be on a diurnal or seasonal basis.

SEASONAL ENERGY STORAGE

In most climate zones, seasonal swings of temperature require buildings to be heated in winter and cooled in the summer. In summer excess heat has to be taken out of the building, and in winter additional heat has to be added to the building. The abundant mass of ground presents an excellent opportunity to store the excess heat of summer to be reused in winter. In certain geographical locations, underground bodies of water such as aquifers may be available, which can also perform this function. In both ground storage and aquifer storage, very precise energy balances are required. Excess use of either heating or cooling over time can lead to accumulation of either heat or cool, which will lose the effectiveness of the storage medium.

Large aquifers for smaller buildings will dissipate all the heat, as water is a good conductor of heat, and underground currents can disperse the heat. However, the aquifers can produce a stable source of water medium that is generally around 55°F. Water at 55°F is a better medium to transfer heat in both summer and winter, where the temperature can be as extreme as 10°F in winter and 95°F in summer. When such systems are used, the only energy required is for the process of transferring heat; the energy required to produce heat is completely eliminated. This leads to reduction in the burning of fossil fuels and reduction of greenhouse gases or carbon dioxide.

The heating, ventilating, and air conditioning (HVAC) equipment used inside a building is similar to geothermal heat systems. Geothermal HVAC equipment is

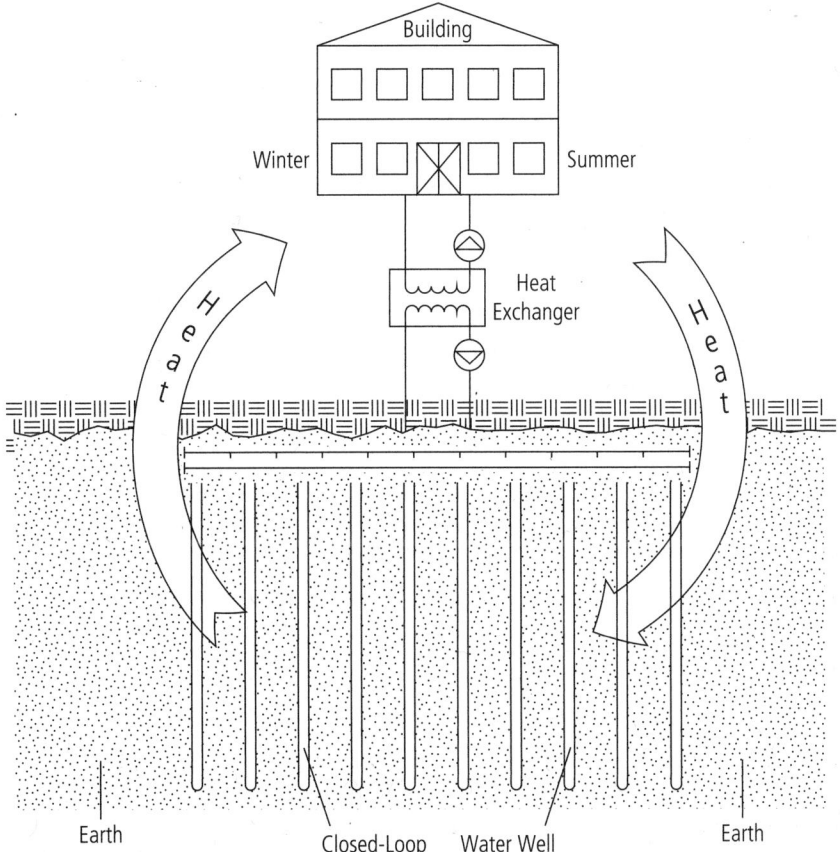

Figure 8–1 Seasonal energy storage.   *Asif Syed*

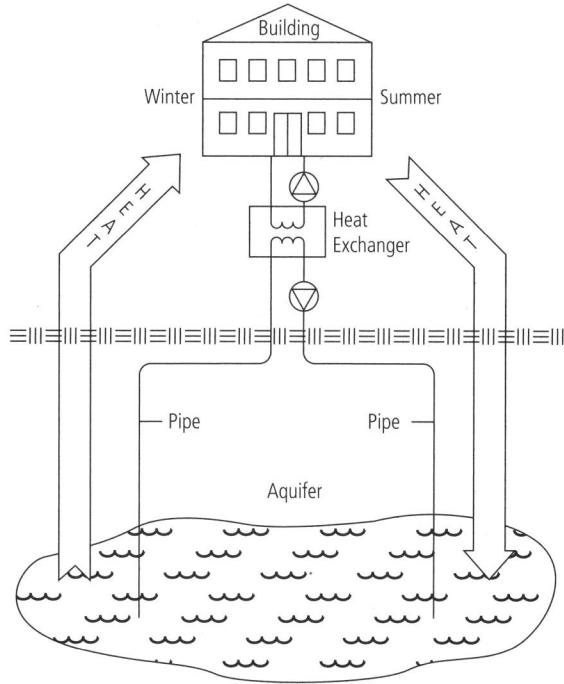

Figure 8–2 Aquifer energy storage. *Asif Syed*

readily available in the market, and contractors are familiar with the systems and their installation. The size of buildings that can use seasonal energy storage is limited by the land available for geothermal storage. A rule of thumb is: For wells 200 feet deep, a 50,000-square-foot building requires 1 acre of land. For wells 400 feet deep, a 100,000-square-foot building requires 1 acre of land. For both ground and aquifer storage, a geological study has to be conducted by an expert in early phases of the programming and schematic design of the project. The study has to establish the thermal properties of the soil and the water and its suitability for the application.

DIURNAL ENERGY STORAGE

Diurnal energy is obtained from the daily variations of temperature and solar intensity. Almost all climate zones exhibit some form of daily fluctuations in temperature and solar intensity. In summer, the nighttime temperatures are much cooler than the day; this is more predominant in desert climates. A typical desert can have temperature fluctuations of as high as 50°F. Low-lying humid climatic zones have lesser temperature difference in the range of 10 to 15°F. In winter the solar intensity during the day is high enough to use it in the night. Diurnal energy is available in two forms: daytime solar radiation and night cool. The daytime solar energy can be stored to heat the building at night. The night cool can be used to cool the building in the day.

SOLAR RADIATION

The most common way to harness the diurnal benefit of solar radiation is the Trombe wall, an architectural feature well known to architects and the construction industry. The technology was patented in 1881 by Edward Morse, and was popularized by Felix Trombe, and named after him. The most common system of the Trombe wall has a thick, solid wall with sufficient mass to store heat; the wall is usually masonry or stone, facing south in the northern hemisphere to catch the solar radiation. The heat storage capacity of the wall is calculated in Btu/hr/cubic feet/°F. An insulated glass double-pane wall with an air cavity between the glass and the mass wall traps the heat in the air cavity (see Figure 8–3). The double-pane insulated glass lets the solar radiation in, but prevents heat loss from the air cavity. During the night, the wall transfers heat to the interior cool spaces. The double-pane insulated glass wall prevents the loss of heat to the exterior. In order to avoid overheating of the Trombe wall in summer an external shade is provided to protect it from the high summer sun, but allows low sun in winter.

Trombe walls can be used in present-day buildings. However, their application has to be carefully evaluated based on the occupancy and use of the buildings. Most commercial office buildings shut off during the night with zero occupancy. In these buildings the nighttime temperature is set back to lower than the daytime occupied temperature. Some heating energy is still required to maintain this lower temperature.

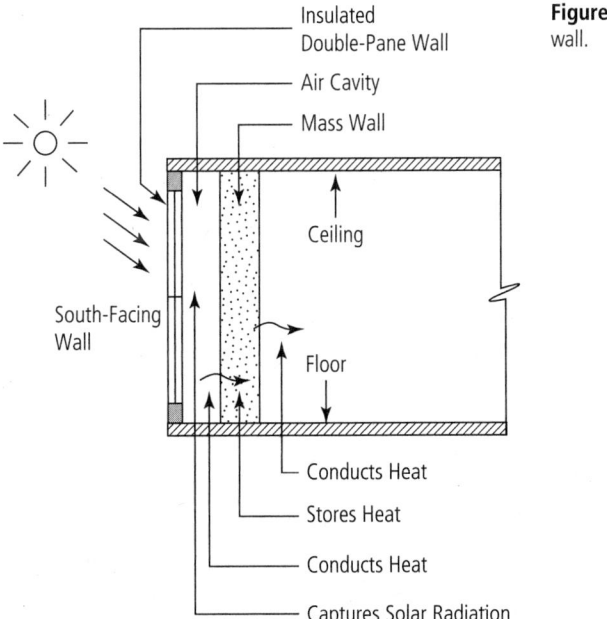

Figure 8–3 Trombe wall. *Asif Syed*

The Trombe wall can be sized to this application. There are several applications where the building continues to be occupied in the night such as student center buildings on college campuses, a train or bus station, or a large public space. In these buildings the energy required to maintain occupied space temperature is much higher and Trombe walls are more effective here. The quality of heat from a Trombe wall is superior due to its radiant effect. The conventional systems heat the air drying it out of humidity.

Modern HVAC systems can work on a similar principle to harvest solar radiation, but instead of storing the solar radiation energy in the mass wall, these systems use it to heat the rest of the building. This is the modified version or one of the several variations of the Trombe wall. Most commercial office buildings do not have night occupancy; the heat captured from solar radiation can be used during the daytime. The heat captured from solar radiation is especially useful in buildings with high ventilation loads, such as schools and research laboratories. The most common application is a double-wall façade. The double-wall façade heat is captured in the air medium, the air is heated, and the heated air is transported to the air-handling unit. Thus, the amount of energy needed to run the air-handling unit—whether steam or hot water or electricity or natural gas—is reduced. In summer, when there is no use for heat, the double-wall façade helps reduce the cooling load by reducing the heat transfer by conduction. There are four different ways a double-wall façade can be used:

1. Outdoor air is circulated to reduce the load.
2. Outdoor air is heated and introduced into the building.
3. Indoor air is circulated to capture heat.
4. Excess indoor air is heated and exhausted, reducing building load.

NIGHT COOL

Night cool is the opposite of daytime solar radiation. At night temperatures drop due to lack of solar radiation on the Earth's surface and heat radiation back into space. The heat gain during the day is continuously lost during the night, until the sun is

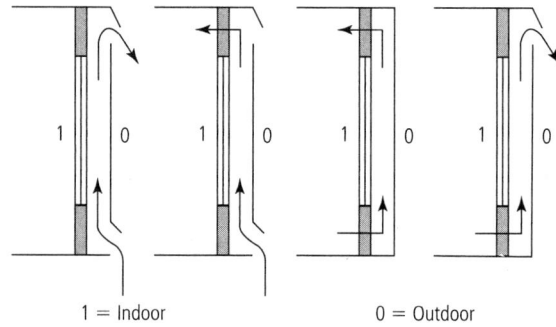

Figure 8–4 Double-wall façades and solar radiation. *Asif Syed*

1 = Indoor   0 = Outdoor

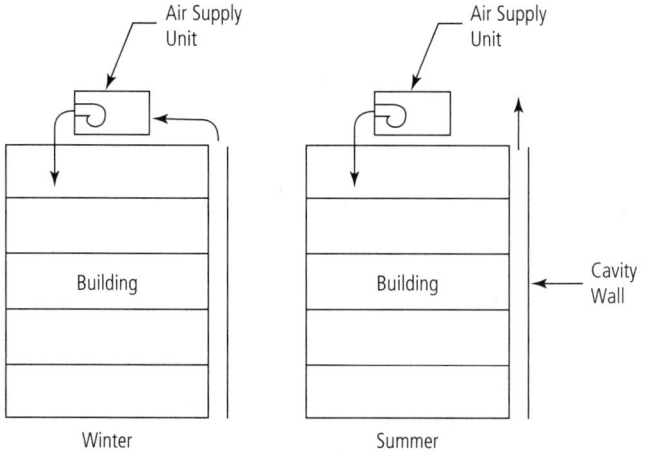

Figure 8–5 Summer and winter use of double-wall façade.   *Asif Syed*

back with its inbound heat radiation exceeding the outbound heat radiation. This is the main reason for nighttime cooling; however, there are localized factors such as wind, jet streams, and so forth, which may have an effect for a few days only. Just as the Trombe wall can store heat, a mechanism can be used in buildings to store cool in the night, to be used during the day. Diurnal temperature change occurs in summer in all climate zones. In climate zones 2 (hot) and B (dry), the temperature fluctuation between day and night is much higher than in moderate climate zones. The temperature fluctuation of Las Vegas and New York is shown in Tables 8–1 and 8–2, respectively.

Tables 8–1 and 8–2 show that average temperature change in Las Vegas is about 25°F, and in New York is about 15°F. By engaging the building mass with the external low temperature of the nights, the building can be pre-cooled for the next day. This reduces the energy required for cooling. There are no ready-made products or equipment available to utilize this diurnal temperature benefit. Lack of ready-made products or equipment makes it harder for design teams to adopt this system in buildings; however, there are opportunities to do so. An integrated approach among architects, structural engineers, and mechanical engineers can engage the building mass with the "night cool."

Examples of Using "Night Cool" to Cool Buildings

The easiest way to take advantage of the diurnal temperature benefit to cool a building in the night is to use the existing air conditioning systems. Most conventional air conditioning systems can provide some means of cooling a building in the night. The effectiveness of the system's thermal diurnal storage has to be verified specific to the site and the building. Diurnal thermal energy storage is especially suitable in desert climates. Evaluation of the sites can present opportunities to save energy and the

TABLE 8–1 LAS VEGAS AVERAGE HIGH AND LOW TEMPERATURE

Location: Las Vegas			
Month	Avg. Low F	Avg. High F	Delta T
Jan	37	57	20
Feb	41	63	22
Mar	47	69	22
Apr	54	78	24
May	63	88	25
Jun	72	99	27
Jul	78	104	26
Aug	77	102	25
Sep	69	94	25
Oct	57	81	24
Nov	44	66	22
Dec	37	57	20

TABLE 8–2 NEW YORK AVERAGE HIGH AND LOW TEMPERATURE

Location: New York			
Month	Avg. Low F	Avg. High F	Delta T
Jan	23	36	13
Feb	24	40	16
Mar	32	48	16
Apr	42	58	16
May	53	68	15
Jun	63	77	14
Jul	68	83	15
Aug	66	81	15
Sep	58	74	16
Oct	47	63	16
Nov	38	52	14
Dec	28	42	14

general thinking that one solution fits all usually does not work. While cooling the building during the night, there are some checks to perform to make sure that the savings are real. These verifications can be made in the design phase. The energy saved by diurnal energy has to be accurate by comparing it with the energy consumed in operating the fans. Fan energy is significant in central air systems, and most fans in building HVAC systems are only 65 percent efficient. Also fans add about 1 to 2°F heat to the circulating air; this heat is generated by the inefficiency of the fan. Condensation must be avoided by making sure that the temperature of the concrete cooled in the night is above the dew point of the space during the morning start-up and operation of the building during day time.

CONVENTIONAL AIR CONDITIONING SYSTEMS

In conventional air conditioning systems, the central air-handling equipment cools and circulates air to the space. The air-handling units have ducted systems that supply air to the space through insulated ducts. The air outlets are located in the ceilings. The space above the ceilings creates a plenum in which ducts and other building services are installed. Air is returned to the air-handling unit through the ceiling plenum, in most commercial building applications. This arrangement presents an opportunity. The plenum is sandwiched between a floor or the roof slab and the ceiling. When the

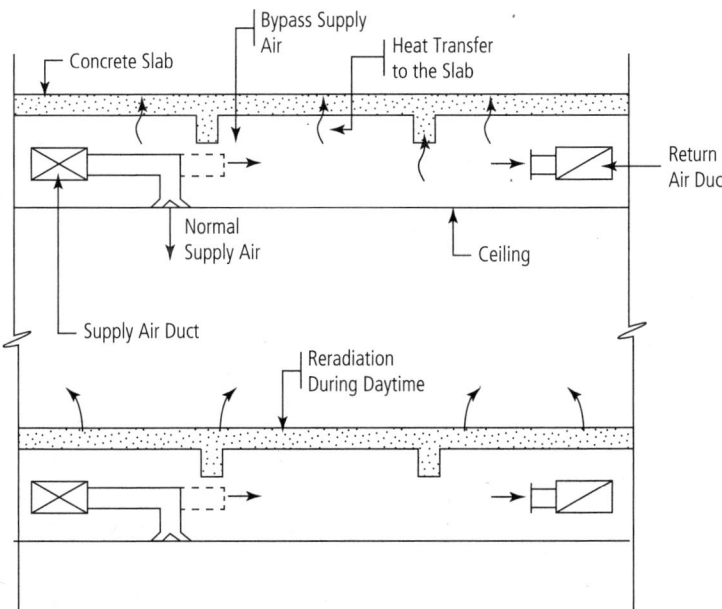

Figure 8–6 Diurnal thermal storage with conventional air conditioning system. *Asif Syed*

floor slab is concrete, it provides mass to store thermal energy. In concrete buildings, the floor slab is in direct contact with the circulating air. In most steel buildings with a metal deck, there is a fire proof insulation, which reduces the heat transfer. The return air plenum is the space between the floor slab and the ceiling. There is opportunity to engage the floor slab with the cold outside air in the night. The air that is normally supplied to the air outlets through the ducts in the ceiling plenum can be bypassed to the return air, creating a short circuit air path through the ceiling plenum making air come in contact with the floor slab, thus cooling the plenum. In turn, the plenum, which is in direct contact with the concrete slab, cools the concrete slab—the largest mass element in the building. Although this system provides easy cooling, it comes with some challenges. Humidification control is a challenge. The cold air in the night may have a dew point that can cause condensation. The humidity or possible condensation challenge can be overcome with control strategies. The outdoor and indoor humidity, and the dew point temperatures indoors and outdoors, have to be monitored to avoid a condition in which condensation can occur.

UNDERFLOOR AIR DISTRIBUTION (UFAD)

Underfloor air distribution (UFAD) systems present the best opportunity and one of the most effective ways to use night cool to cool the building. Underfloor air distribution systems have two air plenums. UFAD systems are more effective than conventional systems for this purpose, because the circulating air comes in contact directly with the floor slab, which is generally made out of concrete, and has a good mass and is suitable for thermal inertia. Even in buildings with metal decks with concrete topping, concrete is in direct contact with the circulating air. All UFAD systems have a supply air plenum at the floor level and a return air plenum at the ceiling. Some UFAD buildings may not have a return air plenum or ceiling, and this does not diminish the effect of storing night cool. Supply air in UFAD systems is normally sent through a central air-handling unit system that cools and distributes air. The supply air is then delivered through a raised floor plenum, which also acts as a supply air plenum. There are no ducts to individual floor air outlets. One duct from the central air-handling unit discharges air into the plenum. This provides an opportunity to engage the floor slab directly with cool nighttime outside air. For UFAD systems, ductwork system modifications are not required. During nighttime, most UFAD applications, which are typically office buildings, are not occupied. During this time when the outside air is colder than the indoor daytime temperature, cold air is supplied to cool the concrete slab and exhausted outdoors. In order to reduce the fan energy, the air system is operated at partial capacity. At lower air flow rates the duct losses are much lower and minimize the fan horsepower. The fan electrical power is proportional to the cube of the fan speed. At one-third the fan speed the power is reduced to one-ninth.

CONVENTIONAL AIR CONDITIONING SYSTEMS 155

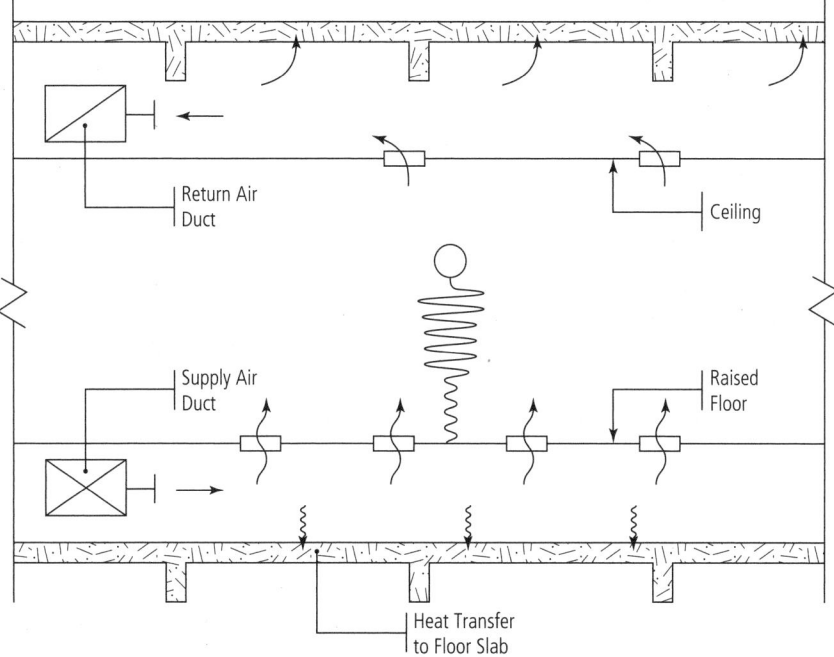

Figure 8–7 Diurnal thermal storage with UFAD systems.   *Asif Syed*

When UFAD systems are operated at night, building management system control strategies must be adopted to overcome potential problems with humidity and condensation. Condensation can occur by cooling the concrete slab below the dew point temperature during the early hours of start-up or during the regular operating hours.

USING COOLING TOWERS

Most large buildings have central air conditioning systems with cooling towers that reject heat outdoors by cooling the water circulated through them. In commercial office occupancy buildings the cooling towers normally operate during the day and are shut down at night or operate at partial load, because buildings are empty. Even in buildings that have nighttime occupancy such as hotels and residential buildings there is no solar load in the night and cooling towers operate at partial capacity. The idling cooling tower capacity in the night can be used to transfer diurnal energy to the concrete mass of the building. The hydronic system of the cooling tower can be used to circulate cold water generated in the cooling tower due to the cold nighttime temperature. Water can be circulated through the building's concrete floor slabs by a tubing system similar to a radiant heating and cooling system. The tubes are made up of nonmetallic and nonferrous materials. The tubes are manufactured in long lengths

156 THERMAL ENERGY STORAGE

Figure 8–8 Diurnal thermal storage using cooling tower. *Asif Syed*

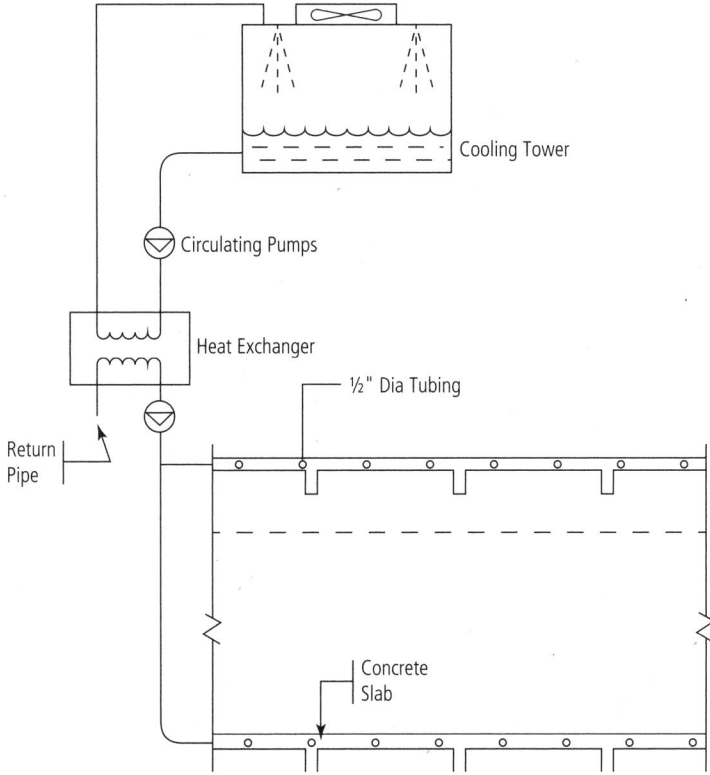

(up to 2,000 feet) that will prevent joints in the system. The tubes are embedded in concrete floors and circulate water. During the night, when the temperatures drop, the cooling tower becomes more efficient and cools water to much lower temperatures than during the day. The fan energy to circulate air is replaced with cooling tower and pump energy. Water has higher specific heat and pumps are far more efficient than fan systems.

NONRENEWABLE ENERGY STORAGE

Nonrenewable energy is generated mostly from burning fossil fuel. It can be produced anytime when it is required. Why produce and store nonrenewable energy for later use? The reason is electricity demand and electricity production and distribution infrastructure. The energy demand is not constant; it varies daily and with seasons. In a 24-hour period, the peak demand occurs for a limited window of time during the day and during the night the demand drops off. The nighttime demand is referred to as off-peak. The cost of infrastructure to fill the peak demand is very high. The production and distribution infrastructure of energy for peak demand requires investment that has a long payback, due to its limited use. Investors don't like long paybacks, so peak electricity is expensive. In order to reduce the payback time on investments, the plants that are designed to provide peak power are usually low-cost and not environmentally as friendly as the base-load plants. The peak load plants may burn fossil fuel such as coal or oil, whereas the base-load plants may be nuclear or hydro or natural gas. The carbon footprint of nuclear and hydro is negligible compared to that of fossil fuel plants. The natural gas power plants are more efficient than coal and fuel oil. Real-time carbon release per unit of electricity (KWh) is a true measurement of carbon intensity. Real-time carbon measurement measures instantaneous carbon release averaged from the plants that are operating on the grid at that time. The benefit of such measurement is that it provides data to consumers to change their behavior as to when to use electricity. These decisions can be as simple as operating dishwashers at night, to operating chiller plants at night to produce thermal energy to be stored and used later during the day. Real-time carbon intensity as measured in the United Kingdom can vary by 25 percent.[4] The cost of electric power at peak hours is high and may sometimes be four times that of nighttime or off-peak or base-load use. Therefore, nonrenewable energy storage offers benefits both to the environment and economics. The option to store electrical energy is limited to batteries. Charging electric car batteries at night has a lower carbon intensity than during daytime peak hours. The most effective way to store this nonrenewable energy in buildings is in the form of thermal energy. Electric energy can be converted into thermal energy and stored. Storage of thermal energy offers additional benefits to the environment by using the diurnal temperature benefit. At night the ambient temperatures are lower, and chillers operate at higher efficiency at lower ambient temperatures than at the higher ambient temperatures of the daytime. The benefits of energy storage are:

1. Lower operating cost of electricity
2. Lower carbon footprint, due to lower real-time carbon
3. Lower carbon footprint, due to lower ambient operating temperature at nights

THERMAL ENERGY STORAGE

The most common form of thermal storage system in buildings is the chilled water or ice storage system. Most buildings, except for hospitals, hotels, and residences, operate only during the day; at night their loads drop to a negligible level. In buildings with 24/7 occupancy, the load drops off at night due to lack of solar load and drop in ambient temperatures. At nights the building air conditioning equipment is idling

---

[4] Real Time Carbon (realtimecarbon.org), AMEE UK Ltd.

or operating at partial capacity. This is the reason that plenty of electrical capacity is available, and the costs are low. The air conditioning load in the building is also not constant for the occupied hours. The air conditioning load peaks in the early afternoon hours and is lower during the early morning hours and later afternoon hours. This presents an opportunity to size the chiller plant at the average load for 24-hour operation, store energy in the form of chilled water or ice, and discharge the energy as required. Thermal storage can be partial or full load. The partial systems are more common; these systems store thermal energy only to offset the peak loads or the spike in the load at peak hours. This results in a smaller storage system. The full-load systems, on the other hand, can store the thermal energy for the entire day. These systems are not as common in buildings, because they require far more space. However, the full-load thermal storage systems can be found in district cooling plants, which serve multiple buildings.

Table 8–3 demonstrates how chillers can be operated at night and chilled water used during the day. This table is only to demonstrate the system operation and not meant to show the efficiency of the system.

Types of Thermal Storage Systems

The most common types of thermal storage systems are:

1. Cold or chilled water storage
2. Ice storage
3. Ice and water slurry storage
4. Eutectic salt storage

The chilled water storage system stores water at temperatures of 40 +/− 2°F. This temperature is the same as or slightly lower than that of the normal conventional

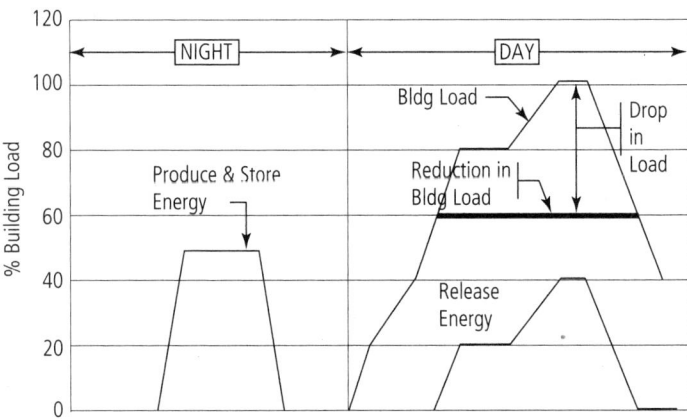

Figure 8–9 Nonrenewable energy: partial thermal storage system. *Asif Syed*

TABLE 8-3 PARTIAL STORAGE SYSTEM OPERATION

Hrs	Day/Night	Mode	Bldg Load %	Store/Release %	Net Load %
19	Night	Produce energy and store	0	0	0
20			0	0	0
21			0	0	0
22			0	0	0
23			0	50	0
24			0	50	0
1			0	50	0
2			0	50	0
3			0	0	0
4			0	0	0
5			0	0	0
6	Day	Discharge from storage	20	0	20
7			30	0	20
8			40	0	40
9			60	0	60
10			80	20	60
11			80	20	60
12			80	20	60
13			100	40	60
14			100	40	60
15			100	40	60
16			80	20	60
17			60	0	60
18			40	0	40

chilled water system. Chilled water storage systems do not involve any phase change (ice to water). This eliminates the benefit of latent heat of fusion. The latent heat of fusion is 144 Btu/lb. The specific heat of water is only 1 Btu/lb or for every drop in temperature of 1°F, the energy stored in 1 pound of water is 1 Btu. The latent heat of fusion presents an opportunity to store more heat for the same volume of water. The chilled water storage systems are larger than ice storage systems. The large-volume chilled water tanks can be located outdoors. Ice storage systems are generally located inside buildings, in the basement, but they can also be located outdoors and on roofs. Ice storage systems have a penalty of efficiency loss. Making ice, which is at a temperature of less than 32°F, reduces the efficiency of the chillers. Ice storage systems also introduce secondary coolant systems and isolation heat exchangers. The secondary coolants are generally glycol solutions which do not freeze at water-freezing or ice-making temperatures. Eutectic salt systems try to achieve a balance between the pros and cons of chilled water and ice storage systems. Eutectic salt solutions freeze at a much higher temperature than ice. The freezing temperature of eutectic salt solutions is 45°F, which is close to the conventional chilled water temperature. The eutectic salts have the advantage of latent heat of fusion. The latent heat of fusion of the eutectic salts solution is 41 Btu/lb, which is lower than that of water, 144 Btu/lb.

Figure 8–10 Outdoor chilled water storage tank. *Asif Syed*

The benefit from lower latent heat of fusion of eutectic salt solutions is still beneficial compared to chilled water storage. The eutectic salt solutions are contained in small plastic containers, and do not mix with the chilled water system. The eutectic salt systems do not have the penalty of chiller efficiency loss of ice systems, and also eliminate the need of secondary coolants and heat exchangers.

SPACE REQUIRED

For all thermal storage systems, space is required to store ice or chilled water. As discussed in the section "Types of Thermal Storage Systems;" ice storage systems require less space than chilled water systems. The storage tanks can be located indoors or outdoors, depending on the availability of space. Availability of space in the building is an important factor in determining the feasibility of a particular thermal storage system. The cost of construction of the space or the loss of rental revenue for the space must be factored into the economic analysis of the system. New York City offers a good example of the importance of space for thermal storage systems. The electric cost can be an average of five times higher than the base rate of 10 cents per kilowatt hour.[5] The higher cost at peak demand times of the day easily justifies the thermal storage system. However, the cost of building the space or of lost rental revenue has made them less attractive. In most buildings where the cost of real estate to rent or build is not high, thermal storage can be easily accommodated. Roofs, basements, and the outdoors can be used to install the thermal storage systems.

---

[5] David Cay Johnston, "Taking Control Of Electric Bill, Hour By Hour," *New York Times*, January 8, 2007.

CHAPTER 9

Solar Energy and Net-Zero Buildings

THE TOPICS OF SOLAR ENERGY AND NET-ZERO BUILDINGS ARE COMBINED together, because solar energy is the primary energy source for most net-zero buildings. Solar energy is abundantly available at almost all places on the Earth, and is easily harvestable on-site at buildings. Photovoltaic (PV) modules covering about 0.3 percent of the land area of the United States, or one-fourth of the land occupied by roads, can supply all the electricity consumed in the United States.[1] There is an abundant amount of solar energy available to be harvested and used in buildings. Net-zero buildings can adopt sources other than solar energy, such as geothermal, tidal, wind, biomass, and so forth, to harvest energy on-site or off-site. However, the most common form is solar energy, and therefore this chapter focuses on solar energy and net-zero buildings. Buildings consume about 40 percent of the total energy consumed in the United States.[2] Net-zero buildings can make a significant impact lowering the energy demand. Net-zero buildings are gaining popularity due to increasing costs of fossil fuels, awareness on the impact of greenhouse gas emissions, and significant progress made on technologies to reduce energy consumption in buildings. Executive Order 13514, signed by President Barak Obama in October 2009, requires all new Federal buildings that are entering the planning process in 2020 or thereafter be "designed to achieve zero-net-energy by 2030." The American Institute of Architects 2030 commitment has targets achieving net-zero by 2030, and the U.S. army has a net-zero program and has selected

---

[1] U.S. Department of Energy, Solar Energy Technologies Program (SETP).
[2] U.S. Department of Energy, Annual Energy Review, 2006.

more than seventeen posts including the West Point Academy for pilot net-zero-installations.[3] In addition, the California Energy Commissions has recommended adjusting the California energy code to adopt net-zero performance in residential buildings by 2020 and commercial buildings by 2030.[4] There is a strong movement to eventually make all buildings net-zero, but there is very limited knowledge and expertise among the broader building design professionals. Net-zero design deviates significantly from the conventional practices in the building design and construction. An integrated design approach with new technologies is required. There has to be participation from the building owner and users as well as a commitment from the design team.

Net-zero buildings can be achieved with a three-step process: (1) harvest solar or other forms of renewable energy, (2) improve the energy efficiency of buildings and their mechanical and electrical systems, and (3) reduce consumption by eliminating waste. All three aspects of net-zero buildings will be discussed in this chapter. In addition to the three steps, integrated design approach is essential and it is a combined effort of all stakeholders in the building.

There is abundant solar energy on the Earth. In some locations there are plenty of other forms of energy, such as wind, geothermal, and tidal. Depending on the site of the building, the most available forms of energy have to be evaluated and adopted. Solar energy is the most common form of energy available in almost all locations, and annually the Earth receives about 3,850,000 exajoules (EJ).[5] All the nonrenewable resources contained within the Earth, such as fossil fuels and nuclear energy from mined uranium, are small compared to the solar energy reaching the surface of the Earth. Solar energy has always been part of the lives of human beings on Earth; ancient homes in all climates of the world either harvested solar energy to keep warm or avoided solar energy to reduce heat gain. The modern technological method of harvesting solar energy dates back to even before the First World War. In 1913, Frank Shuman, a native of Pennsylvania, built a solar thermal collector system in Egypt, to produce steam from solar reflecting mirrors, which in turn pumped water to irrigation sites in the arid desert land. Further development of the project was stopped due to the outbreak of the First World War. After the war, the new oil fields discovered in the Middle East put an end to Frank Shuman's solar energy plan.

Increasing energy efficiency by designing buildings with efficient mechanical and electrical systems has become easier. Several technologies are available to reduce energy consumed by the mechanical and electrical systems, some of which were

---

[3] "Army Selects 17 Posts to Attain 'Net Zero' Consumption," April 2011, www.defensecommunities.org/headlines/army-selects-17-posts-to-attain-%E2%80%98net-zero%E2%80%99-consumption/#.
[4] "California to Require Net-Zero-Energy Buildings," www.buildinggreen.com/auth/article.cfm/2008/2/3/california-to-require-net-zero-energy-buildings/.
[5] Mervin Johns, Hanh-Phuc Le, and Michael Seeman, "Grid-Connected Solar Electronics," University of California at Berkeley, Department of Electrical Engineering and Computer Sciences, EE-290N-3 – Contemporary Energy Issues.

explained in the earlier chapters. The conventional technologies of heating and cooling buildings were developed when energy costs were so low compared to the installation costs that energy efficiency did not matter. Annual energy consumption was not the focus. However, with the present-day high costs of energy, the operating costs of buildings are large enough for building owners to think about energy efficiency. Attention to the effects of greenhouse gases has also brought about a focus on energy use, as most of the energy comes from fossil fuels. Efficiency is getting the focus it deserves. Guidelines and books are available to reduce energy use—for example, *Advanced Energy Design Guide for Small to Medium Office Buildings: 50% Energy Savings*, from the American Society of Heating, Refrigerating and Air-Conditioning Engineers (ASHRAE).

Reducing consumption is also an important aspect of achieving net-zero buildings. The behavior of building occupants is involved in this aspect of reducing consumption of energy. Occupants have to ask themselves whether they really need what they are using. However, this should not lead to a lowering of standards or to leaving legitimate needs unfulfilled. What needs to be controlled is excessive use of energy that is often wasted on unwanted, unused applications. One good example is the television in the living room that does not have to be on when no one is sitting there to watch it. This television can be shut off manually or automatically. Similarly, in office buildings, the computers can be put in sleep mode when users are not at their desks. Private office lights can be turned off with occupancy sensors. Studies have shown that occupant awareness, involvement, and commitment to energy use reduction can make a big difference even in conventional buildings.

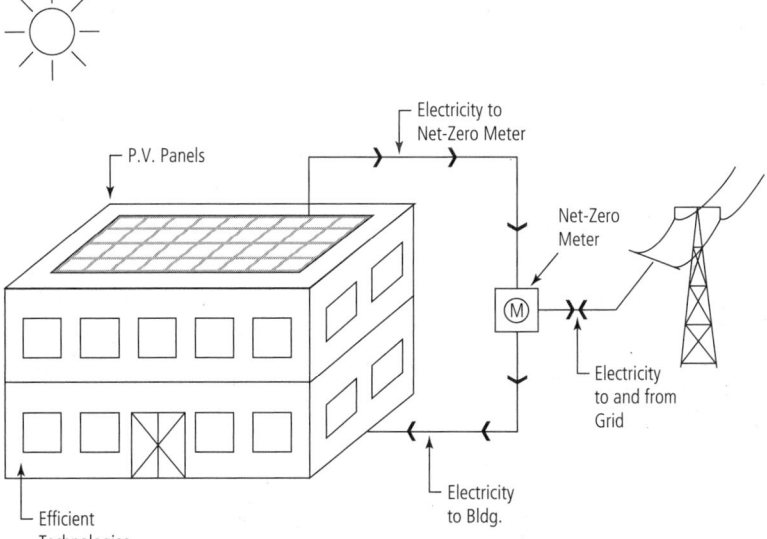

Figure 9–1 Net-zero energy building with solar energy. *Asif Syed*

NET-ZERO STEP 1: HARVESTING SOLAR ENERGY

The surface area of the Earth receives approximately 137 watts per square foot[6] This energy is in the form of visible light, ultraviolet, and infrared; after the reflection of clouds, an average of 100 watts per square foot reaches the Earth's surface. This is an abundant amount of energy when compared to what is used in the buildings. As a rule of thumb most commercial buildings use an average of 6 watts per square foot. The electrical power incoming to the building is generally sized for 6 watts. Some buildings may use higher or lower—it is an average that can be used to demonstrate the potential of solar energy in buildings. If all 100 watts per square feet of solar energy is captured the roof area of a building can support almost fifteen floors of the same area. However, the entire 100 watts per square feet of incident solar energy cannot be captured. The current technologies available limit the amount of useful energy that can be captured. The efficiency of harvesting solar energy varies for heat and electricity generation. Achieving the right mix for a building based on the site, occupancy, and use is a critical factor in net-zero buildings.

EFFICIENCY OF SOLAR PHOTOVOLTAIC CELL

The efficiency of electricity production with solar photovoltaic panels is about 15 percent. Only about 15 percent of the energy received by the photovoltaic panel is converted into electricity. This is similar to power generation at a fossil fuel power plant, where only about 30 percent of the energy in the fossil fuel is converted into electricity, and the remaining 60 percent is lost in the form of the heat of the waste gases. The electricity produced by photovoltaic panels is direct current (DC), whereas the most common form of electricity used by commercial and residential applications is alternating current (AC). There are more losses when DC is converted into AC by devices called inverters; the combined efficiency of conversion is about 75 percent. The actual useful electricity produced for 100 watts per square foot of solar energy on the panel is about 11.25 watts (discounting for 15 percent efficiency of the panel and 70 percent efficiency for DC to AC conversion). For non-commercially-available photovoltaic cells, the efficiency is higher, but these cells are not practical for buildings. Embodied energy of the photovoltaic or solar cells has been a question in the minds of designers. How much energy goes into making the photovoltaic cell? Many believe that energy that goes into making solar photovoltaic cells is almost unrecoverable. The U.S. Department of Energy, Office of Energy Efficiency and Renewable Energy

---

[6] U.S. Department of Energy Astronomy Archives: Solar Radiation on Earth, www.newton.dep.anl.gov/askasci/ast99/ast99413.htm.

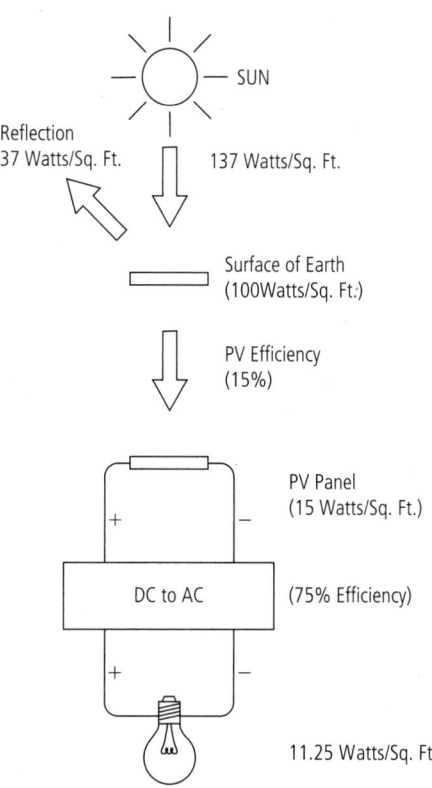

Figure 9–2 Solar photovoltaic energy efficiency. *Asif Syed*

answers this question and clarifies it in their publication, PV Frequently Asked Questions.[7] The energy payback from photovoltaic cells can vary from one to four years, based on the technology of the photovoltaic cell.

HARVESTING SOLAR THERMAL ENERGY

Solar energy captured as heat is commonly known as solar thermal energy. In solar thermal energy, the energy from the sun—in the form of visible light, ultraviolet, and infrared—is converted into heat generally hot water. The efficiency of solar thermal is much higher than that of solar photovoltaic. Its efficiency at capturing solar energy is 60 to 70 percent, or five to six times greater than that of photovoltaic. The solar thermal energy is captured in solar thermal panels, which collect heat to produce mostly hot water in building applications or steam in some large-scale industrial applications. In cold climates, antifreeze is used in solar thermal collectors to prevent freezing in the night. Hot water is commonly used in buildings, for space and domestic heating.

---

[7] U.S. Department of Energy, Office of Energy Efficiency and Renewable Energy answers this question in their publication, PV FAQ, www.nrel.gov/docs/fy04osti/35489.pdf.

In some applications hot water can be used for air conditioning, using the absorption technology. Domestic hot water is needed all year; however, heating is limited to the winter months. This limits the use of solar thermal energy that is available all year round. Solar thermal energy is captured with solar thermal panels. There are different types of solar thermal panels and they are classified according to temperature of hot water generated. The common classification is low temperature, medium temperature, and high temperature.

1. Low-temperature solar thermal collectors generally operate at 5 to 30°F above the ambient. The most common use for these collectors is pool heating. There are industrial and agricultural applications, which are not discussed here. The collectors are nonglazed, or not in a glass enclosure. They are bare metal tubes that are exposed to sunlight and circulate water.

2. Medium-temperature solar thermal collectors generally operate at 15 to 200°F above the ambient temperature. Medium-temperature solar collectors are enclosed in a casing with glass. Casing maximizes the capture and transfer of heat to the water tubes.

3. High-temperature collectors concentrate solar energy by use of parabolic mirrors to produce steam. The application is mostly industrial and generally not found in the building sector.

Flat plate collectors: The most common type of solar thermal collectors are flat plate glazed collectors, with insulated back plates, a black or dark-colored absorber plate to absorb heat, and hot water circulating pipes. The absorber plates are made out of metal and selective coatings are applied to increase the absorption of heat. The insulation at the bottom reduces heat conductance to outside the collector, maximizing the capture of heat. The common use is domestic water heating and building space heating.

Evacuated or vacuum tube collectors: Evacuated or vacuum tube collectors are more efficient than flat plate collectors. The improvement in efficiency comes from the vacuum tubes, which are similar in principle to a thermos bottle. The vacuum helps in retaining the captured energy by reducing convection losses. Vacuum collectors cost more than flat plate collectors, but are generally cost effective in colder climates. Evacuated or vacuum tube collectors are either direct type or heat-pipe type. In direct collectors, circulating water flows directly through the heat-collecting tube. In heat-pipe-type collectors, a phase-change liquid acts as an intermediary between the collector tube and the circulating hot water. The phase-change material evaporates and condenses in the collector tube. The heat transfer rate of heat pipes is much higher than in the conduction and convection process. The evaporative process allows the capture of heat for solar energy to a greater extent, as the latent heat of evaporation of the fluid is much higher. The phase-change or heat-pipe collector has to be

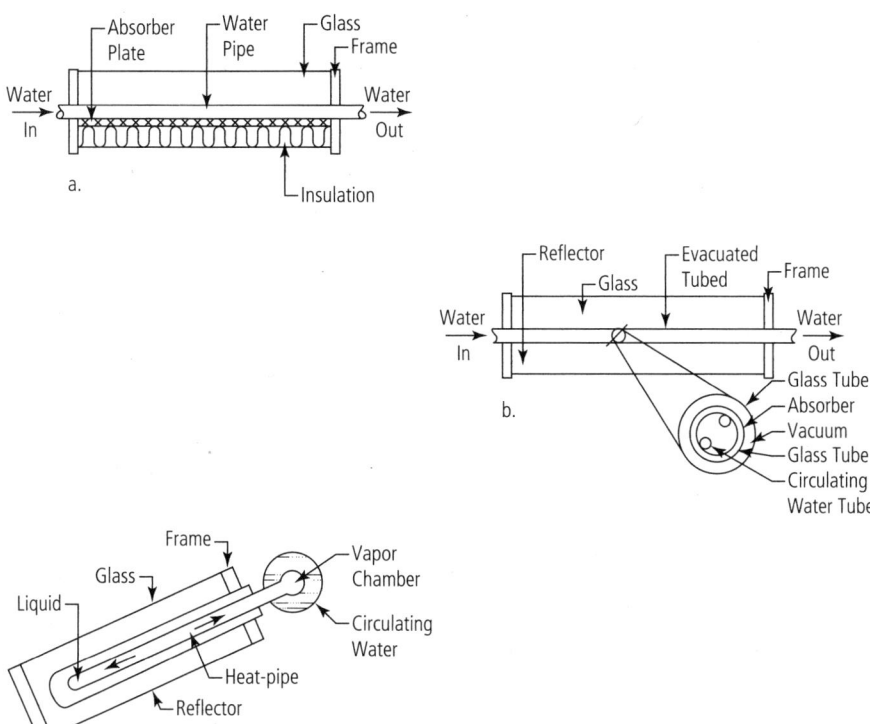

Figure 9–3 a. Flat plate collector. b. Evacuated or vacuum collector. c. Heat-pipe evacuated collector. *Asif Syed*

installed at an angle or tilt. The tilt is required to make the phase-change material flow by gravity in the phase-change tube. However, the tilt can be taken advantage of in most locations to face the sun to achieve the best angle of incidence.

EFFICIENCY OF SOLAR THERMAL COLLECTORS

Efficiency of solar collectors is defined as the ratio of usable thermal energy efficiency to the solar energy received over a unit area. The efficiency of solar collectors is dependent on the type of collector and the temperature of water above the ambient. Based on the type of collector, efficiency improves from flat plate to vacuum or evacuated to heat-pipe collectors. As the temperature difference between the ambient and the collecting medium (such as water) increases, the efficiency drops. This efficiency drop is due to thermal losses, which are higher at higher temperatures. The average efficiency of solar collectors ranges from 60 to 70 percent. For 100 watts of solar energy received on the Earth's surface, about 60 to 70 watts can be captured to heat hot water.

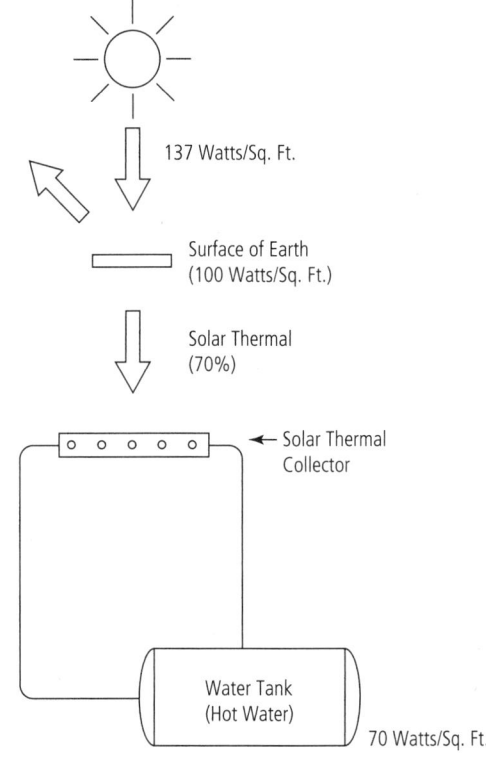

Figure 9–4 Solar thermal energy efficiency. *Asif Syed*

SOLAR PHOTOVOLTAIC VERSUS SOLAR THERMAL

The average efficiency of a photovoltaic system is about 12 percent. The average efficiency of a solar thermal collector system is 70 percent. For the average of 100 watts per square foot solar intensity that reaches the Earth's surface and for the same collector area, solar photovoltaic captures about 12 watts per square foot, and solar thermal captures about 70 watts per square foot. Solar thermal is six times more efficient than solar photovoltaic. To capture the same amount of energy, solar thermal requires one-sixth the area needed for solar photovoltaic panels. Solar thermal produces heat, and photovoltaic produces electricity. Heat and electricity are different forms of energy and have different uses in buildings. The primary use of electricity is in appliances such as refrigerators, washing machines, televisions, air conditioning systems, and lighting. Hot water can be used for domestic water needs, such as in the bathrooms, and for space heating. Hot water can also be used to operate the air conditioning system. The technologies for operating air conditioning systems by solar thermal are limited and expensive, but are available.

TABLE 9–1 DIFFERENCES BETWEEN SOLAR PHOTOVOLTAIC AND SOLAR THERMAL

Photovoltaic System	Solar Thermal System
12.5% efficient	60–70% efficient
12.5 watts per square foot (average)	70 watts per square foot (average)
Energy delivered to the building as electricity	Energy delivered as hot water
Large roof area per unit of energy harvested	Smaller roof area per the same unit of energy harvested
Energy stored in batteries	Energy stored in hot water tanks
Energy easily exported off-site via grid	Limitations to off-site export

Storage of energy: Both electricity and heat can be stored for nighttime use in buildings, when there is no sunlight. Electricity is stored in batteries, and hot water is stored in insulated hot water tanks. Another form of solar energy storage for nighttime use is hydrogen gas. Hydrogen gas is produced by electrolysis of water during the daytime. Hydrogen can be stored to be used in the night and on cloudy days. Fuel cell technologies have advanced to the use of hydrogen to produce heat and electricity on demand. However, the production of hydrogen via electrolysis of water for commercial applications is not fully developed. In both commercial and residential buildings, solar thermal use is generally limited to domestic hot water, space heating, and air conditioning. Photovoltaic-generated electricity is easily transported off-site, because it is connected to the electric grid.

Export of energy off-site: Electricity produced by a photovoltaic system in excess of what is consumed at the site can easily be exported to the grid system. Most buildings are connected to an electric grid, with a robust infrastructure of transmission lines, transformers, and power plants. North America has three major grids: (1) the Western Interconnection, (2) the Eastern Interconnection, and (3) the Electric Reliability Council of Texas (ERCOT). There are limitations as to how far solar-generated power can be transmitted over the grid. However, power can be exported within reasonable distances to areas where other consumers can use the power. The ease with which photovoltaic electricity can be exported to off-site gives it a big advantage. Additionally, the peak production of solar power generated electricity coincides with the peak demand for electrical power. The greatest demand for electrical power is during daytime hours, the same time when solar energy is produced. Solar thermal, on the other hand, has limitations on export and demand. There are almost no hot water distribution grids. Therefore, solar thermal energy cannot be exported off-site on a large scale, as with electricity. However, on small-scale applications hot water distribution infrastructure can be easily built, such as on college campuses, where roof areas on administrative buildings can be used to harvest hot water, and it can be transported from there to residence or dormitory buildings. Domestic hot water use is accurately predictable and is generally constant for a 24-hour cycle and 365 days a year, and this facilitates the accurate estimation of energy required. Roof areas for highly efficient use of solar thermal energy for the domestic hot water can be estimated. The hot water use for the building's space-heating is seasonal. During the shoulder seasons between summer and winter, space heating is not required. This limits the maximization of the use of solar thermal energy in buildings. The six-fold advantage of solar thermal in energy-capture efficiency has a disadvantage that it cannot be used throughout the year. Solar thermal collectors designed for winter use do not operate at full capacity in summer. This wastes the valuable roof area in the summer. Therefore, there is a right mix of solar thermal and photovoltaic, which balances the advantages of both forms of solar energy capture, use, storage, and export.

The right mix: Every building has the right mix of solar thermal and photovoltaic. A larger solar thermal system is very efficient and costs less, but is idle for many hours in a year, due to lack of demand. A smaller solar thermal system reduces the potential to maximize the capture of energy from the sun. The solar photovoltaic system has lower efficiency and captures less energy from the sun, but there is always demand for all the energy produced. When the demand in the building drops, energy can always be exported to the grid. Every building's right mix of electricity and solar thermal depends on a detailed analysis of annual energy use (broken down into hot water and electricity); the roof area, solar panels array area, or thermal collector area; and storage and export capabilities. There is no one solution that fits all buildings. The complexity of the evaluation and analysis of the building is an important factor as to why these are not more common. Such analysis is iterative; there is no ready-made software available that can find the optimum mix. The analysis has to be performed by a skilled and knowledgeable engineer. The cost of the engineering time is relatively high, and most small projects cannot afford to do this analysis; they therefore lose out on the potential for energy savings from solar energy. Most state governments are encouraging such building energy use analysis by cost sharing with building owners and developers.

SOLAR INTENSITY

Solar intensity, or direct normal solar radiation, is the amount of energy received at a particular location. Solar intensity is measured in kilowatt hours per square foot per day or kilowatt hours per square meters per day. It is averaged over the useful hours of sunlight, a period of about six hours. During the early morning and late afternoon hours, the angle of incidence is too low for useful radiation. One square foot of area receives 100 watts of solar radiation; averaged over a six-hour period, this equals 600

TABLE 9-2 AVERAGE ANNUAL SOLAR INTENSITY FOR DIFFERENT LOCATIONS IN THE UNITED STATES

State	City	County	Zone Definition	Solar Intensity kWh/m^2/day	Solar Intensity kWh/ft^2/day
Florida	Miami	Dade	1A	5.17	0.48
Florida	Orlando	Orange	2A	5.23	0.49
Georgia	Atlanta	Fulton and DeKalb	3A	4.99	0.46
District of Columbia	Washington	District of Columbia	4A	4.67	0.43
New York	New York	New York	4A	4.49	0.42
Illinois	Chicago	Cook and DuPage	5A	4.51	0.42
California	Los Angeles	Los Angeles	3B	5.54	0.51
Washington	Seattle	King	4C	3.67	0.34
Arizona	Phoenix	Maricopa	3B	6.18	0.57

watts or 0.6 kilowatt hours, although the actual numbers vary based on the location. For the climatic zones in the United States, the average solar intensity can vary from 0.48 to 0.57 kilowatt hours per square foot per day. The National Renewable Energy Laboratory (NREL) has developed a software program, PVWatts, that calculates the solar intensity of every geographic location worldwide. The calculator is available on the Internet and is free of charge. In addition to providing solar intensity, PVWatts also calculates total energy monthly and annually, enters the cost of electricity based on a database, and calculates the cost of power that is generated by photovoltaic at the selected location. These are costs that are avoided by not purchasing the electric power from the utility. Table 9–2 provides the average annual solar intensity for different locations in the United States. It can be seen from Table 9–2 that almost all locations listed in the table are good for solar energy.

ANNUAL SOLAR ELECTRIC POWER GENERATION

Annual power generation from a solar PV system can be calculated using software that is readily available on the Internet, such as PVWatts, developed by the National Renewable Energy Laboratory (NREL). The rule of thumb for power generation at a given location is:

Solar intensity or direct solar radiation
0.51 kWh/ft^2/day (e.g., Los Angeles)

TABLE 9-3 RULE OF THUMB FOR PHOTOVOLTAIC POWER GENERATION

Area ft²	x	Solar intensity kWh/ft²/day	x	365 days	x	PV cell efficiency	x	DC to AC efficiency	=	Annual kWh
6,600		0.42		365		0.15		0.77		116,865

Multiply by 365 days in a year (cloudy and rainy days not accounted for)
0.51 × 365 = 186.15 kWh/ft²/year
Multiply by photovoltaic cell efficiency (example 15%)
186.15 × .15 = 29.72 kWh/ft²/year
Multiply by DC to AC conversion efficiency (example 77%)
27.92 × .77 = 21.50 kWh/ft²/year

An example of a commercial building located in New York would be as follows: Say the roof area available is 6,600 square feet. With a 15 watt-per-square-foot rated PV system (6,600 ft² × 15 W/ft² = 99,000 W = 99 kW), a 99-kilowatt system can be installed. For a New York location, the solar intensity is 0.42 kilowatt hours per square foot per day. It is recommended that the software be used to calculate the solar annual power generation.

The result from PVWatts software for the same location and a 100-kilowatt system is 119,781 kilowatt hours. The detailed printout from the software appears as shown in Table 9–4.

BUILDING ENERGY CONSUMPTION

Net-zero buildings are designed to generate as much energy as they consume annually. This is a quantitative measure. In order to understand the science behind the design of net-zero buildings, both architects and engineers need a sound understanding of the building energy consumption. The splits in energy consumption between heating and cooling and the occupants' use, also known as plug loads, need to be understood. As the energy consumption reduces from adoption of advanced technologies, the plug loads used by occupants become important. Building energy consumption is defined as the annual energy consumed in the building. The number is usually in kBtu/sq. ft./year. The number has two definitions: (1) energy measured at site and (2) energy measured at source.

Site energy is the energy consumed at the site boundaries or physical lot line of the land plot on which the building is located. A good measure of site energy consumption is the sum of all the electric bills or kWh (electrical energy units) and the sum of natural gas bills or therms (energy unit of natural gas). Both these measurements are relatively easy—just the addition of utility bills. Therms and kWh are two different units of energy; both units are converted into the common unit of kBtu.

TABLE 9-4 PRINTOUT FROM PVWATTS SOFTWARE

Station Identification		Results			
Cell ID:	269370		Solar	AC	Energy
State:	New York	Month	Radiation	Energy	Value
Latitude:	40.8° N		(kWh/m²/day)	(kWh)	($)
Longitude:	73.7° W	1	3.08	7610	1332.97
PV System Specifications		2	3.92	8701	1524.07
DC Rating:	100.0 kW	3	4.95	11676	2045.17
DC to AC Derate Factor:	0.77	4	5.03	11089	1942.35
AC Rating:	77.0 kW	5	5.35	11927	2089.13
Array Type:	Fixed Tilt	6	5.57	11681	2046.04
Array Tilt:	40.8°	7	5.21	11030	1932.01
Array Azimuth:	180.0°	8	5.16	11031	1932.19
Energy Specifications		9	5.04	10703	1874.74
Cost of Electricity:	17.5 ¢/kWh	10	4.43	10105	1769.99
		11	3.19	7205	1262.03
		12	2.90	7024	1230.32
		Year	4.49	119781	20980.84

Source energy tracks the energy to the origin or point of generation of energy. The power plant is the source of origin for electricity. At the power plant, the energy source is either fossil fuel, such as coal, oil, or natural gas; or nuclear fuel; or hydro. The energy that is used to produce electricity is the embodied energy of the fuel released at the power plant. Generally the source energy is higher than site energy due to losses in production and transportation. For example, the electricity produced at a thermal fossil fuel power plant is generally only 30 to 40 percent efficient. The rest of the energy is lost as heat in the flue gases. For all forms of energy there is a site to source conversion factor which will be discussed separately. The average building delivered energy consumption, or site energy consumption, figures are:[8]

---

[8] U.S. Department of Energy, Buildings Energy Data Book: 3.1 "Commercial Sector Energy Consumption," U.S. DoE, 2010.

TABLE 9–5 AVERAGE SITE ENERGY CONSUMPTION FOR BUILDINGS

No.	Type of Building	Btu/sq. ft./year
1	Residential	58.7 (1)
2	Commercial — office buildings	88 (2)
3	Health care (hospital)	253.8 (2)
4	Educational	80.6 (2)

(1) Average of all regions 2005 residential delivered energy consumption intensity
(2) Average of 1990 to 2003 commercial delivered energy consumption by building type

SOURCE-TO-SITE ENERGY CONVERSION FACTOR

Understanding of the site-to-source conversion factor is vital to the science of energy consumption in buildings. This concept also helps to understand and compare use of different forms of energy with a common benchmark and equitable units of measurement. Source energy consumption is the only measure of carbon footprint of any energy-consuming process. The source energy figure accounts for the energy of the fuel at the plant plus the transmission losses (over the power lines for electricity) from the power plant to the point of use. Source energy consumption is significantly higher than the site energy for electricity. For electricity, the source energy used is three times as much as the site energy, with fossil-fuel-burning power plants. The reason for high source energy consumption is twofold: (1) loss due to transmission and (2) power generation losses. The transmission losses depend on the distance from the plant to the building. The power generation losses depend on the technology of power generation. Most fossil fuel power plants have an efficiency of 30 percent. Only 30 percent of the energy of the fossil fuel consumed is generated into electricity, and 60 percent goes out as waste heat in the flue stacks. The term "site-to-source conversion factor" is commonly used to define the ratio of site energy to source energy. The conversion factor depends on the type of fuel used at the power plant. Hydro power plants do not use fossil fuel and do not have power generation losses. The hydro plant losses are limited to transmission losses. Therefore, hydro power plants have a lower site-to-source conversion factor. The U.S. Environmental Protection Agency (EPA) has established that source energy is the most equitable way of evaluating the energy consumed by a building.[9] The conversion factors depend on the region,

---

[9] U.S. Environmental Protection Agency, ENERGY STAR Performance Ratings: Methodology for Incorporating Source Energy Use, March 2011.

TABLE 9–6 SOURCE-TO-SITE CONVERSION FACTORS—EPA'S NATIONAL AVERAGE

Fuel Type	Source-to-Site Conversion Factor	Remarks
Electricity from grid	3.34	Generation losses (3.04 – 3.02) and transmission and distribution losses (1.2 – 1.31). Average = 3.34.
Electricity on site (PV or wind)	1.0	Generation losses are zero, as no fossil fuel is involved. Transmission losses are zero, as power is produced at site.
Natural gas	1.047	The losses account for pipe distribution and transmission.
Fuel oil (diesel)	1.01	The losses account for trucked distribution, storage, and dispensing.

the type of fuel consumed, and the equipment used. The conversion factor differs from region to region. Some areas of the country use more fossil fuel, such as coal, and other areas have hydroelectric power plants. The EPA has developed a program called Portfolio Manager to measure building energy consumption and building efficiency. Portfolio Manager uses a national average conversion factor to evaluate or rate buildings for their energy efficiency. The national average system does not disadvantage a building for being located in a region with a lower conversion factor. The national average conversion factors, as published by the EPA, are shown in Table 9–6.

SOLAR ENERGY IN NET-ZERO BUILDINGS

At peak loads, most buildings consume less than the useful energy that is captured from the 100 watts per square foot of solar intensity. The average useful energy captured with a photovoltaic system at peak solar conditions is 12.5 watts per square foot. The average energy that can be captured from solar thermal is 70 watts per square foot. Assuming 20 percent solar thermal and 80 percent photovoltaic, the average energy that can be captured is 24 watts per square foot. With an average of six hours of useful solar conditions, the total solar energy that can be captured is 150 kBtu/sq. ft./year per square foot of sunlit area. This even accounts for some overcast days of lower solar energy. As seen in Table 9–5 earlier, the average energy consumption of buildings is 58.7 kBtu/sq. ft./year for residential buildings and 88 kBtu/sq. ft/year for commercial buildings. Almost two to three times the amount of solar energy is available per square foot of building interior area as per square foot of roof area. Therefore for a two- or three-story building, there is sufficient energy available from the sun to meet the energy consumed in the buildings to make them net-zero. The above equation can be used only as a demonstrative technique to show how buildings can be self-sufficient. It does not mean that all two- or three-story buildings can become self-sufficient; in real-life buildings there are several challenges to overcome to make the above equation work. Some of the challenges are that supply and demand times do not match, storage of energy is expensive, and there are high capital costs. There are several different ways that these challenges can be overcome and all of them are unique to the building, occupancy, climate zone, return on investment, and so on. There is no one net-zero solution that fits all buildings. There are several small steps that can be taken in the design of buildings and each step leads closer to the net-zero. These steps include reducing energy consumption in buildings with passive methods to heat, cool and light, use advanced technologies to reduce energy consumption of mechanical and electrical systems, tap into other renewable resources such as geothermal, and so on. However, it is important to note that there is abundant solar energy available to harvest and use in buildings.

Net-zero building design teams need to do a detailed analysis in early phases of design and adopt an integrated approach with participation of all stakeholders. The approach identified in the equations above here is only meant to demonstrate that net-zero energy buildings are achievable. It is by no means a replacement for a real detailed analysis of an actual building, whose results may or may not agree with this approach. Detailed analysis is required for every building, based on its location, occupancy, energy use intensity, and the type of energy used. The detailed analysis usually requires an annual energy simulation of the building. This energy simulation includes useful input parameters such as:

1. The simulation based on a 365-day, hour-by-hour basis.
2. Weather data for the building site on an hour-by-hour basis.
3. The software has to be iterative-based.
4. The building envelope parameters must be accurately simulated.
5. The building energy use parameters, such as lighting, plug power, and so forth have to be accurately defined.
6. The occupancy schedules.
7. The equipment operating schedules.

Net-zero buildings are feasible for the majority of the buildings in the United States and in other countries, which are one- or two- or three-story buildings. The ratio of a building's floor area to its roof area is an important factor in the feasibility of achieving net-zero. Net-zero is generally feasible for a building with a floor area to roof area ratio of 2 to 3. For high-rise multistory buildings in urban environments, on-site net-zero is generally not feasible, because they do not have sufficient roof space. Generally for single- and two-story buildings a combination of solar power thermal and photovoltaic systems can produce more or as much power as they can use. The excess power can be transported into the grid, which in turn can support the energy needs of other buildings which cannot be net-zero such as high-rise buildings.

NET-ZERO DEFINITION

The general definition of a net-zero building is a building that produces as much energy as it consumes. Another term commonly used for net-zero buildings is zero energy buildings (ZEB). The measure of energy consumed differs depending on how energy is defined by different parties involved with the buildings. These parties could be the owners who pay the electricity bills, the utility company that supplies electricity and natural gas, or the government agencies that formulate policies. The three different parties have different interests. "Energy" may mean the cost of the utility bills to a homeowner or commercial building owner. Reduction in energy bills may be the most important criterion for the owner. However, carbon emissions are most important to the environmentalist, because energy use generates greenhouse gases and contributes to global warming. But to a government agency "energy" may mean the embodied energy of the fuel consumed at the site. The varying interests of the different parties involved pose practical problems in the formation of a common definition. This has led to several definitions for a net-zero building or zero energy building, and to confusion in the building design community, as to which definition to use. It is important to establish or identify the definition that will be used by the design team. This decision has to be made at the very start of the project. All parties concerned or associated with the net-zero goal must be assembled, in the early phases of project design. All the different approaches that lead to different definitions must be discussed, and it is important to establish or select a definition that meets the goals of the stakeholders. The three major definitions of net-zero are:

1. Energy (Btu or kilo calories)
2. Cost in dollars (cost of energy)
3. Emissions (carbon dioxide or greenhouse gases)

The three major indices can be further subdivided into the boundary condition, or the extent of the boundaries within which they are measured. The boundaries can be local or global.

1. Local or site (boundary line of the building plot)
2. Global or source (extended boundary to power plant or energy source)

Net Metering

Most utilities are required by law in the states where they are operating to purchase photovoltaic or other renewable or cogeneration electricity generated in buildings, when it is in excess of what the buildings use. Net metering provides a mechanism to measure the electricity in and the electricity out, at a meter. Incoming electricity is from the utility power company, and outgoing electricity is generated at the site, in excess of what is consumed. The electricity exported can be generated during the day from a photovoltaic source, and electricity can be consumed during the night when there is no solar energy. A wind power station may produce power day or night when there is wind; this power may be in excess of what is consumed at the site, due to time of the day or the wind speed, and this excess may be exported to the grid. The goal of net metering is to account for the export, and to ensure that the reduction in the electric bill to the consumer equals the outflow of electrical power from the building. The Energy Policy Act of 2005 mandated that utilities buy back renewable power generated by their customers' buildings and provide a net metering methodology for metering electrical power. In most renewable power-generating systems in buildings, at any given instant the energy produced and the energy consumed do not match. In some instances, all the energy produced will be consumed; for example, in an office building with an electric air conditioning system, in the middle of a weekday, the electricity demands of lights and air conditioning will consume all solar photovoltaic energy produced. In other situations, such as weekends in an office building, the electricity produced will be in excess of what is consumed. Similarly, in winter, if the building uses gas heat the electricity produced may be in excess of what is consumed. In residential buildings, the potential of solar photovoltaic is well in excess of what can be consumed, due to the large roof area and the minimal use, as most homes are not occupied during daytime hours. Therefore, there is excess energy in some buildings and a shortage in others, creating a need to export the excess. Excess energy produced from renewable resources is particularly beneficial. This energy does not use fossil fuels associated with greenhouse gases or nuclear fuel associated with radioactive waste and power plant disasters, such as the damage caused by the tsunami in Japan in 2011.

Net metering allows electricity to be purchased back at the same price at which it is sold to the customer by the utility. The term "net meter" means that the same meter measures the current in the reverse direction and balances out incoming and outgoing energy. A basic electric meter can easily net out the incoming and outgoing current. In real life, net metering is not as simple, because the cost of electricity is based on several factors, not just on the cost of energy. The price of electricity is generally broken down into the cost of power (commodity charge), the time of day of use (demand charge), and the cost of distribution (distribution charge). The cost of power is the cost of fuel and the power plant, and the cost of distribution is the cost of the infrastructure of the power lines and network and the cost of losses in transmission. The cost of power is high during peak demand hours of the day, and it costs less during the night hours; similarly, the cost of distribution mimics the cost of power, as there is excess distribution capacity at nights. In a net-zero program, the utility companies may pay the same price for the commodity and demand charge, but not pay distribution charge. The distribution network is still owned by the utility or ownership of distribution infrastructure may be different. Therefore, net metering does not necessary mean net-zero cost. The tariff structure is different for different utility companies. Prior to designing a renewable power-generating source in a building, it is important to evaluate the available tariff structure for net metering. Generally, a computer simulation of annual power generation, export, import, and consumption, with the tariffs from utility companies plugged in, will determine the optimum size of the solar photovoltaic systems. Solar photovoltaic systems are expensive, therefore their optimum cost benefit needs to be evaluated along with the environmental benefits.

NET-ZERO STEP 2: IMPROVE ENERGY EFFICIENCY OF THE BUILDING AND ITS MECHANICAL AND ELECTRICAL SYSTEMS

Improving energy efficiency and reducing energy consumption requires an aggressive analysis of energy efficiency measures. Step 2, improving energy efficiency of the building, helps in step 1, harvesting renewable energy. Less energy consumed equates to less energy needed to be harvested. In order to reduce energy consumption, a benchmark for measurement is required. The most common and universally understood benchmark is the energy code ASHRAE 90.1. The goal of step 2 is to reduce energy consumption compared to the energy code. Step 2 is the process of reducing the size of the energy pie. With currently available technologies, there are significant opportunities to reduce the energy consumption of buildings. Net-zero buildings can reduce their energy consumption from 50 to 75 percent from the benchmark energy code. Reducing energy consumption by 50 to 70 percent is doable; however, there is no one solution that fits all buildings. Each building requires unique energy efficiency measures. Therefore understanding the science behind the energy reduction, conservation, efficiency improvement, and recycling or reusing of wasted energy, is important in achieving the net-zero. In order to make a significant reduction in the energy used by a building and its mechanical and electrical systems, a holistic approach is required. Individual efforts by different members of the team, such as architects and engineers, separately working toward the best possible technologies, will not work. Most energy-efficient technologies that make a significant savings in energy have an impact on more than one discipline. For example, displacement ventilation or underfloor air distribution (UFAD)—both these mechanical air distribution systems have an impact on both building interiors and architecture. Similarly, building orientation—an architectural discipline—can be optimized by a building energy simulation performed by a mechanical engineer.

It is also important to study the financial aspects of the project such as the cost of the energy efficiency measures and their payback, each measure at a time. When multiple energy saving measures are lumped together the individual value of each measure is lost. Collectively the sum of the measures may meet the financial objectives, some measures may have too long of a payback and could be eliminated. This helps in identifying the cost benefit of each energy saving measure. For some measures the synergies between each other will increase or multiply the energy savings. Most building owners and developers have an investment payback or return on investment goal. Energy simulation coupled with capital costs, life-cycle cost analysis and simple cost-versus-savings payback calculations are required. This assures that the project stays in budget. Inaccurate information on capital costs of measures and payback can lead to big problems later, which will be difficult to solve due to collective or integrated approach of the designs. Problems that arise cannot be fixed by individual trades, the collective solutions will take more time and may delay the schedule. The utility rate structures have to be factored in the cost and payback analysis. The utilities rate structures can be complex, with multiple options to purchase utilities, varying from fixed pricing to market rates to options and futures. The building owners and operators have to be educated about the options for them to purchase utilities. The carbon or greenhouse gas emissions of the different utilities have to be accounted for and calculated for different combinations of the utilities. User input is required to account for the end use of the building, which is a critical factor in determining the net-zero energy plan. This complexity of issues leads to the holistic or integrated design approach. The integrated design approach helps in early collaboration and involvement of all stakeholders. A clear and achievable goal can be established only with all stakeholders agreeing to work toward that goal.

All possible technologies have to be considered in addition to ones discussed in this book. The technologies have to be evaluated and analyzed specifically for the project site conditions. All technologies don't work for all climatic zones—for example, chilled beams may not work in high-humidity regions; therefore, site-specific evaluation is required. A top list of technologies, not listed in any particular order, that can generally work toward net-zero are:

1. High-performance envelope
2. Radiant heating and cooling/Chilled beams/Displacement induction units
3. Displacement ventilation
4. Thermal storage

5. Geothermal
6. Cogeneration
7. Lighting: daylighting harvesting and controls
8. Passive cooling and natural ventilation
9. Energy recovery

NET-ZERO STEP 3: REDUCE CONSUMPTION

To reduce energy consumption:
1. Optimize the use of energy (turn off equipment when not in use).
2. Use the most energy-efficient equipment (LED TV vs. LCD).
3. Encourage design professional interaction with the end user.
4. Do not overdesign.
5. Correlate the cost of energy use with the cost of renewable energy production.
6. Empower the end user with knowledge of energy use.

Reducing energy consumption should not mean using less energy by sacrificing or lowering standards. Reducing energy consumption means utilizing energy in an optimum way, to maximize its benefits. A good example is turning the television off when leaving the room. Thus, reducing energy use does not mean not watching television, or reducing the time spent watching it. Similarly, an office computer monitor and desktop computer can hibernate when the user is not directly working with the device.

The next way to reduce energy consumption is to use the most energy-efficient equipment. The example of television can be extended to the types of televisions available or to the technologies available. Choose the television with the lowest energy consumption. For example, LED technology uses much less power than LCD. Use ENERGY STAR–labeled equipment. ENERGY STAR is a program jointly developed by the U.S. Environmental Protection Agency and the U.S. Department of Energy. The ENERGY STAR program labels equipment and appliances that strictly meet the energy efficiency guidelines set by both agencies.

Design professional interaction with the end user is the best way to understand how equipment is used and energy is consumed. Design professionals may not necessarily be familiar with or understand the habits or behavioral aspects of the end user. However, having the end user engaged in the design aspects creates a lot of value. For the television example, shutting off the television may mean different things to different people. Some may choose to install a wall switch to turn off while leaving the room. Others may want to default the television to turn off after a preset time.

Figuring out the end users and providing them with a design feature that enables their preferred method of shutting off equipment will create an excellent opportunity to reduce energy consumption. However simple this appears, it can save a lot of energy in energy-intensive projects such as laboratory buildings. Fume-hood areas use extensive amounts of once-through heated and air-conditioned air. Some hoods, such as teaching or demonstration hoods, need not be exhausted at the full rate when chemicals are not stored in the hood. Storing the chemicals separately can significantly reduce the air change rate and energy consumption. The normal practice of the design professional is to assume that all fume hoods with storage cabinets will store chemicals. However, interacting or engaging with the end users provides the design professional with insight into the use of the equipment, and provides a design that is energy efficient.

Overdesign happens when designers follow preset industry standards without understanding the end user. An example of overdesign is the office building HVAC system. An industry standard generally accepted by developers, tenants, and engineers for cooling plus lighting equipment heat load is 5 watts per square foot. Most buildings rarely realize this, and in recent years office lighting technology and office computer equipment have become more energy efficient. A larger HVAC system designed for 5 watts per square foot may be excessive, resulting in equipment not operating at its peak efficiency. Actual use can be established only via interaction with the end user.

Making the energy use data visible to the end user significantly impacts energy use behavior. Easily observable and accessible energy consumption information increases the occupant's energy consciousness and creates behavior change. This behavior is similar to stock-trading activity in the financial markets—easily available stock portfolio performance information leads to more trading activity. The most common method of displaying energy consumption information is via dashboards. The dashboard displays live energy use information, including intensity of energy use. Such a dashboard in an office will prompt occupants to open blinds and shut off lights at the perimeter.

CHAPTER 10

Geothermal Systems

INTRODUCTION

Geothermal energy is the energy stored in our planet since its formation. There are four different forms of energy below the outer crust of the Earth: (1) heat left over from the condensing of the planet from gases, 4 billion years ago, (2) frictional heat from the gravitational sorting or displacements of denser and lighter parts in the liquid core, (3) latent heat of condensing of magma as the Earth cools down, and (4) energy from the radioactive decay of isotopes. The majority of the geothermal energy comes from the core of the Earth, from the radioactive decay process. Radioactivity is present not only in the Earth's core but also on the crust. Chris Marone, professor of geosciences at Pennsylvania State University, explains that 1 kilogram of granite on the surface of the Earth emanates a tiny but measurable amount of heat through radioactive decay.[1] The crust of the Earth has a very limited thickness of 10 to 15 kilometers under the ocean and 35 to 70 kilometers under the continents. Geothermal energy is abundant, but there are no visible signs of it on the surface of the Earth, except in volcanically active areas. Geothermal energy can be used in buildings in several ways, depending on the quality of energy available at the site of a building. The quality is considered good when the temperature of the energy source is high, and poor when the temperature of the energy source is low. In volcanically active areas near the joints of the tectonic plates, geothermal energy is of high quality.

---

[1] Joe Anuta, "Probing Question: What heats the Earth's core?" *Research/Penn State* magazine, March 27, 2006 (http://live.psu.edu/story/16903).

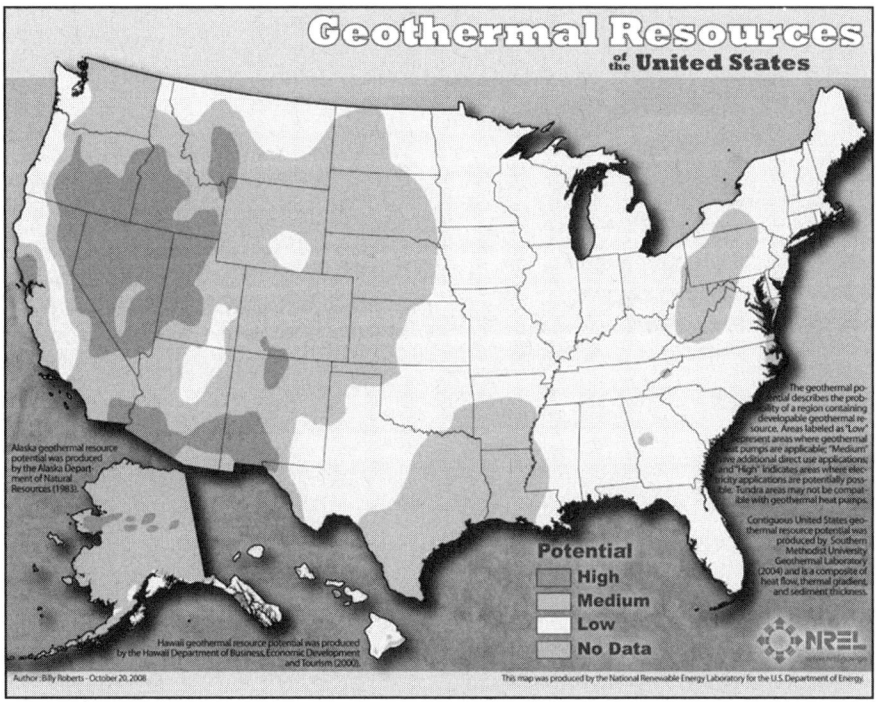

Figure 10–1 Geothermal resource map of the United States. *National Renewable Energy Laboratory*

In the remaining areas—almost the entire surface of the Earth—low-quality energy is available. Just a few feet below the surface of the Earth, the temperature is constant at 55°F, giving us an abundant source of low-quality heat. Detailed geothermal resource maps for all regions of the United States are accessible and available from the National Renewable Energy Laboratory (NREL) of the U.S. Department of Energy. All geothermal energy is free of carbon dioxide emissions because it does not burn any fossil fuel, except for the transfer process, making it a green and sustainable energy source. The carbon dioxide emissions from a geothermal heat pump are the lowest, compared with other heating and cooling systems.[2]

The benefits of geothermal heat pumps can be categorized into three types: environmental, building, and cost.[3]

Environmental benefits:

- Higher efficiency leads to lower electric power consumption.
- Geothermal heat produce zero carbon emissions, which helps reduce global warming (except the energy for transfer of heat).

---

[2] Richard C. Niess, "Geothermal heat pump systems are 'red hot' but are they really 'green'"? The Second Stockton International Geothermal Conference, March 16 and 17, 1998, The Richard Stockton College of New Jersey.

[3] International Ground Source Heat Pump Association.

- Geothermal energy is renewable; heat is extracted from or sunk into the Earth, which maintains constant temperature.
- Geothermal systems eliminate the water use in cooling towers of conventional systems.
- The gas utility to the building can be eliminated, reducing operating costs.
- Heat transfer between cooled and heated areas within the building recycles heat generated in the building.

Building benefits:
- Smaller units provide opportunity for multiple-zone control.
- Simultaneous heating and cooling of different parts of the building is possible without boiler operation or a four-pipe system installation.
- There is reduction in fan energy of large central systems.
- Reduced duct sizes can be used, allowing for higher ceilings.
- Smaller mechanical rooms allow for higher floor-area efficiency.

Economic benefits:
- Lower consumption of electricity and elimination of fossil fuel for heating reduce operating utility costs.
- In one study, over the life cycle of thirty years the life-cycle costs were less by 16 percent, compared to rooftop packaged DX systems.[4] The initial capital cost of geothermal systems is generally higher than that of standard rooftop packaged systems; however, the savings in energy costs over the life of the building lower the total cost of ownership of the building.
- Incentives are available from state and federal programs to offset some of the additional costs of the systems.
- The geothermal system reduces the electric peak demand on the utility, avoiding expensive costs of building power plants.

HEAT FROM GEOTHERMAL ENERGY

The quality of heat energy from geothermal sources is classified as high, medium, or low, depending on the temperature, as follows:

- High grade—Higher than 212°F
- Medium grade—Less than 212°F
- Low grade—About 55°F

---

[4] Andrew Chiasson, "Final Report: Life-Cycle Cost Study of a Geothermal Heat Pump System, BIA Office Building, Winnebago, NE," Geo-Heat Center, Oregon Institute of Technology, February 2006.

High-grade energy can be used to generate electricity in a geothermal power plant. High-grade geothermal energy is abundant in the Earth, but only that which is available at the surface can be used. Geothermal electric power plants can be found in several countries around the world. The most common ones are in Iceland, Greenland, and the Western United States. California, Hawaii, Nevada, and Utah currently have geothermal power plants. In most of these plants, the high temperature is used to produce steam, which in turn runs the turbines. In the United States, the geothermal capacity is 20,000 megawatts. There is potential to multiply this capacity by 5, given undiscovered resources.[5] The process is carbon-emission-free, as it does not burn any fossil fuel.

The areas with medium-grade energy are limited. The heat from medium grade can be used for heating applications. The limited areas of medium-grade heat may not necessarily be where there is enough population density to make use of it. This is one of the oldest forms of geothermal heat use. Reykjavik, Iceland, has the largest geothermal district heating system, where almost all buildings are heated with geothermal heating.[6]

Figure 10–2 Geothermal heat pump operation—cooling mode.   *Florida Heat Pump Company*

---

[5] Charles Kutscher, "The Status and Future of Geothermal Electric Power," conference paper presented at the American Solar Energy Society Conference, Madison, WI, July 16–21, 2000; NREL/CP-550-28204, August 2000.

[6] Geothermal Education Office, 664 Hilary Drive, Tiburon, CA 94920 (http://geothermal.marin.org/).

Figure 10–3 Geothermal heat pump operation—heating mode. *Florida Heat Pump Company*

Low-grade energy heat is abundant and is available over almost the entire planet. Low-grade energy can be used in buildings, which will be the topic of discussion for the remainder of this chapter. Animals take advantage of the fact that the temperature below the surface of the Earth is stable and steady all year round at 55°F. Mammals that hibernate burrow down or enter a den that maintains a temperature above freezing.[7] Even when the temperatures outdoors are below freezing, the burrows maintain a temperature of 55°F. The temperature of air on the surface of the Earth can vary from 120°F in summer to below zero in winter, but the temperature just a few feet below is a constant 55°F. Geothermal applications use this comfortable subterranean temperature to condition the environment of our buildings. The term "heat pump" is self-explanatory—a heat pump transfers heat from one location to another. In cooling mode, heat is collected from the building and sunk into the Earth. In heating mode, the heat is extracted from the Earth and "sunk" into the building. The medium used for this heat transfer is water. As discussed earlier, water is an efficient carrier of heat energy. The energy needed to pump water between source and sink is relatively low.

---

[7] Ken Muldrew (University of Calgary, Alberta, Canada) and Locksley E. McGann (University of Alberta, Alberta, Canada), Cryobiology—A Short Course, a Web-based textbook, http://people.ucalgary.ca/~kmuldrew/cryo_course/course_outline.html.

GEOTHERMAL HEAT PUMPS

The geothermal heat pump is fundamentally an air conditioner. It has the same components as an air conditioner, such as a compressor, a condenser, an expansion valve, an evaporator, and a fan to circulate air. An additional component in the heat pump is the reversing valve, which facilitates the reversal of the source and the sink. In the standard air conditioner, the heat source is inside the building (indoors), and the heat sink is outside the building (outdoors). Standard air conditioners transfer heat from indoors to outdoors. In a heat pump, the heat sink in summer or cooling mode is outdoors (into the Earth), and in winter the heat sink is indoors (inside the buildings). In heat pumps, the sink and source change according to the season.

WHY GEOTHERMAL HEAT PUMPS ARE EFFICIENT

The efficiency of the heat pump comes from two sources:

1. The heat sink temperature is lower because the heat sink is the Earth. The temperature of Earth is always at 55°F, which is lower than the temperature of outdoor air in the summer, when the outside temperature is 90°F to 115°F and indoor temperature is 75°F. This reduces the effort or the energy required to transfer the heat. It is easier to reject heat at 55°F than 90°F to 115°F.

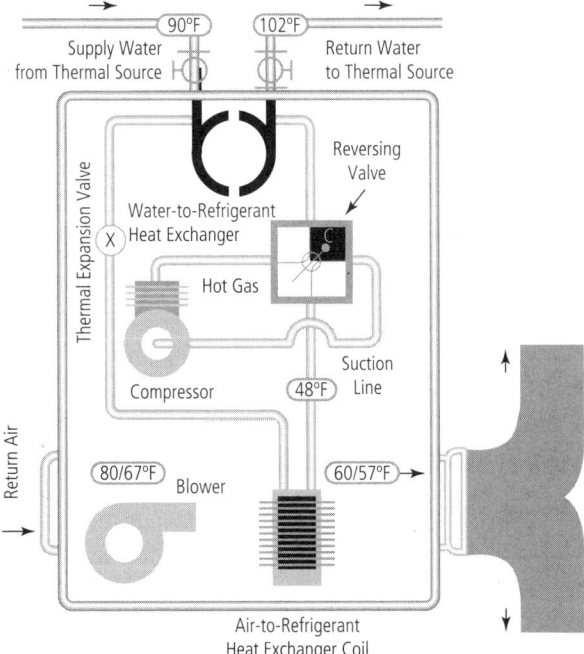

Figure 10–4 Heat pump operation—cooling mode. *Florida Heat Pump Company*

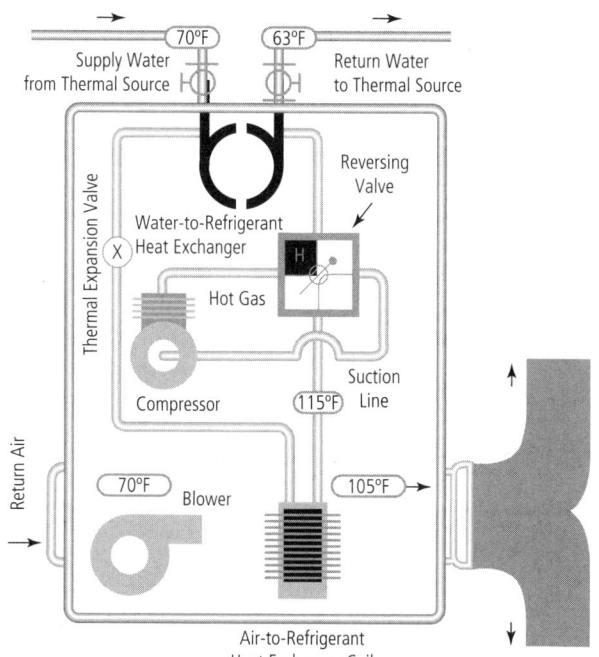

Figure 10–5 Heat pump operation—heating mode. *Florida Heat Pump Company*

2. Free heat is available in the ground. As discussed earlier, the Earth is always at a steady 55°F. In winter, when the outdoors may be at 10°F, some form of fossil fuel is burned to keep the building heated to 68°F, and heat is used to offset the heat losses. The ground source heat pump extracts the heat from the Earth, which is at 55°F, and transfers it to the indoors to offset the heat loss. The energy used is for extraction and transfer of heat. This avoids heat generation by burning a fossil fuel. The level of energy used to transfer heat is lower than the energy used to create heat.

Thermal efficiency of heat pumps is 300 to 500 percent; for every one unit of electricity consumed, three units of heat are introduced in the space. Comparative efficiency of a conventional electric heater is 100 percent; for every one unit of electricity consumed one unit of heat is added to the space. Heat pumps are three times as efficient.

GEOTHERMAL METHODS

The heat extraction and sinking of heat to and from the Earth can be accomplished by several different methods.

- A direct buried closed water pipe loop into the Earth. Here there is no direct contact between water and the soil. Heat is transferred between the building

Figure 10–6 Energy efficiency of geothermal heat pump. *Asif Syed*

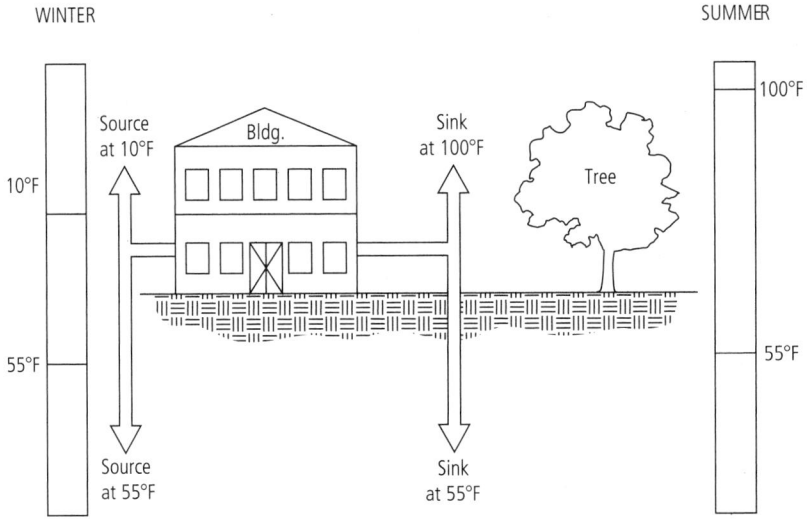

and Earth through a pipe material. Such systems are also referred to as closed-loop systems. Direct buried loops can be horizontal or vertical.

- A lake, pond, or other body of water can be used to harness geothermal energy. The water indirectly transfers heat from the Earth and maintains a steady temperature similar to that of Earth.
- An aquifer or underground lake can be used. The water temperature in the aquifer remains constant. Water is drawn from a well and injected back into the aquifer. These systems are also referred to as open-loop systems.

Geothermal systems are classified into closed loop or open loop. In both systems, water is used as the heat transfer medium between the source and the sink. In a closed-loop system, the heat transfer medium of water does not come in contact directly with the elements of the Earth. Heat transfer takes place through the pipe wall. The risk of contaminants from the Earth entering the piping system is almost zero. Similarly, the Earth does not get contaminated with chemicals used to treat water in the piping system, to inhibit corrosion. Contamination occurs only in case of pipe breaks, which are detectable, and when this happens the wells are put out of service. In an open-loop system, water is extracted from the Earth or a lake or a riverbed and pumped into the building piping loop. After the water is used in the heat pump, it is spilled back into the Earth or the lake or the riverbed. The disadvantage of this system is that the quality of the Earth's water may not be suitable to the building piping system. The water may be acidic or basic, which could be harmful to the pipes. This water is treated with chemicals to make it suitable for use in the piping system. The open loop also has the potential to contaminate the Earth with chemicals used to treat water. There are various types of open- and closed-loop systems (see Table 10–1).

TABLE 10-1 TYPES OF OPEN- AND CLOSED-LOOP SYSTEMS

System	Type	Pros and Cons
Closed loop	1. Vertical 2. Horizontal 3. Slinky 4. Lake or pond	1. Clean water in building piping system 2. No contamination of Earth with chemicals 3. Water temperature in summer is 70°F and in winter 40°F (1)
Open loop	1. Open well 2. Aquifer 3. Lake or pond 4. River	1. Building piping system not protected from water contaminants 2. Some contamination of the Earth 3. Water temperature is steady both summer and winter at 55°F (1)

(1) Water temperatures are averages used to demonstrate; actual temperature may vary from site to site.

VERTICAL CLOSED-LOOP SYSTEM

In a vertical closed-loop system, vertical boreholes are drilled into the ground. The depth of the hole or drilling depends on the quantity of heat to be extracted and the thermal conductivity of the Earth's soil. Generally, the depth of the holes can range from 200 to 600 feet. A pair of pipes with a "U"-joint at the bottom is inserted into the borehole. The boreholes are filled with concrete grout, with the pair of pipes in the center. The concrete grout increases thermal conductivity with good contact with the soil. Vertical closed-loop fields can be located under parking lots, athletic fields, streets, and grassy areas. The piping material is high-density polyethylene. It is recommended that the loop field be identified in a survey with well locations, with a reference to the buildings. Loop identification is important if repairs are needed.[8] The layout of the well farm varies depending on the geology and hydrology of the location. Important properties of the soil such as thermal conductivity and thermal stability must be studied prior to finalizing the design for a geothermal system. For every geothermal project, in addition to a registered professional engineer a certified geothermal designer must be retained. Geothermal designer certification is provided by the Association of Energy Engineers (AEE) with the joint sponsorship of the International Ground Source Heat Pump Association (IGSHPA) and the Geothermal Heat Pump Consortium (GHPC).[9] For preliminary planning purposes, the general rule of thumb is wells at 15 feet on center, 200 feet deep for 1 ton per well, and 400 feet deep for 2 tons per well.

---

[8] Kevin B. McCray, Guidelines for the Construction of Vertical BoreHoles for Closed Loop Heat Pump Systems, National Ground Water Association, Westerville, OH, 1997.
[9] Phil Rawlings, "Earth Insights," *Geo Outlook*, Volume 1, Number 3, Oklahoma State University, Stillwater, OK.

Figure 10–7 Layout and capacity of typical well field. *Asif Syed*

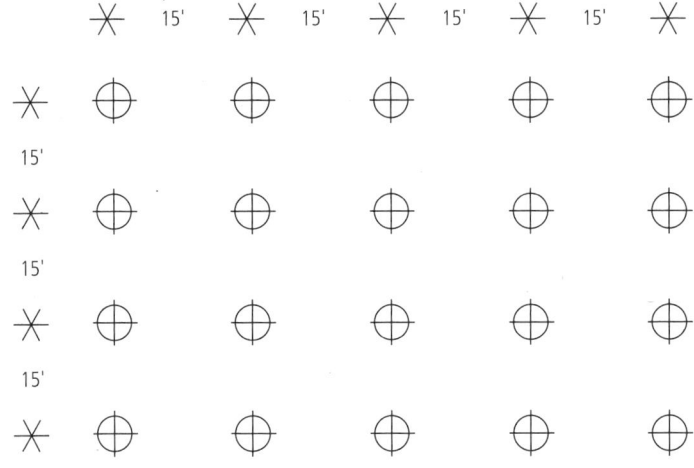

20 Wells	Capacity Each	Capacity Total
200 Feet Deep	1 Ton	20 Tons
400 Feet Deep	2 Tons	40 Tons

HORIZONTAL CLOSED-LOOP SYSTEM

In a horizontal closed-loop system, the pipes are laid horizontally instead of going deep into the Earth. The excavation is limited to a maximum of 10 feet, but the area of the loop farm increases. For a vertical 1-ton, 200-foot-deep well, 225 square feet are required. For the same ton, about 600 to 1,000 feet of horizontal trench needs to be excavated. A 2-foot-wide trench will require an area of 1,200 to 2,000 square feet. The large area requirements make this system suitable for small installations and limit its use in large commercial buildings.

SLINKY CLOSED-LOOP SYSTEM

The slinky horizontal loop is similar to the horizontal loop and is installed in a shallow trench. The trench is usually wider and deeper. The pipes are laid out in a spiral "slinky" coil. The slinky system's advantage is that it reduces the length of the trench. Five-hundred feet of overlapped slinky coils can be placed in an 80-foot trench.[10]

---

[10] *Ground Source Heat Pump Manual*, Commonwealth of Pennsylvania, Department of Environmental Protection, Document No. 383-0300-001, August 23, 2000.

Figure 10–8 Horizontal closed-loop heat pump. *Florida Heat Pump Company*

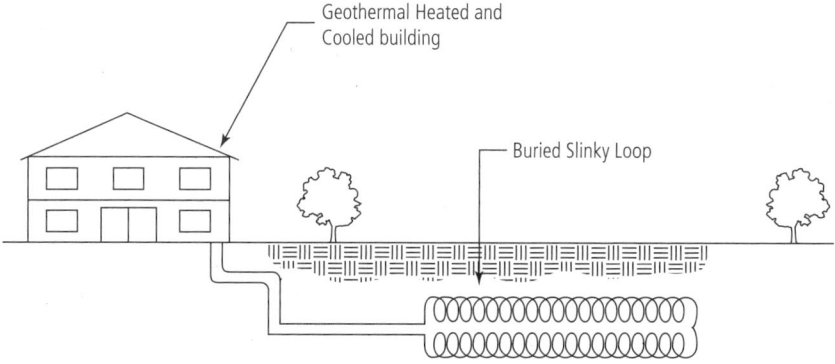

Figure 10–9 Slinky loop system in ground. *Asif Syed*

LAKE OR POND CLOSED-LOOP SYSTEM

Lakes and ponds maintain a higher temperature than air in winter, and a lower temperature in summer, like the Earth a few feet below. The lakes and ponds do not freeze more than 2 to 3 feet in winter, and the water temperature below the freezing

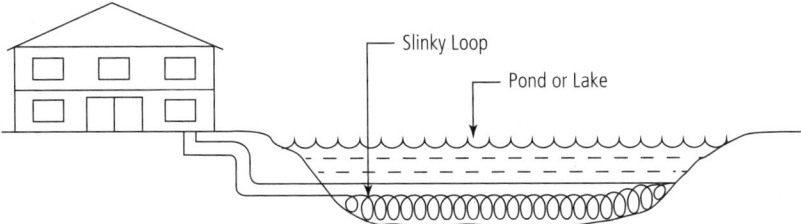

Figure 10–10 Slinky loop system in lake. *Asif Syed*

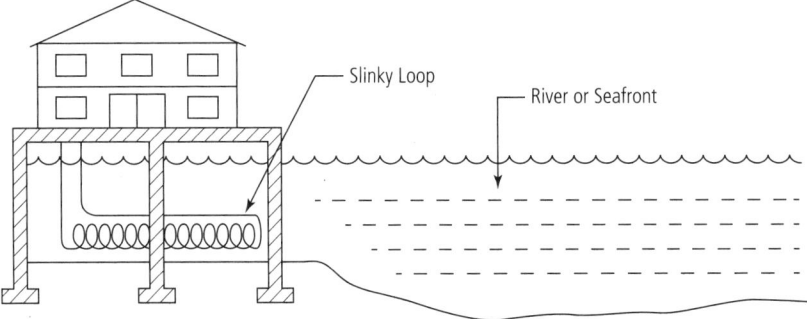

Figure 10–11 Slinky loop in pier. *Asif Syed*

surface allows heat transfer, making lakes and ponds good heat sources and sinks. In a closed-loop system, a slinky coil is laid at the bottom of the lake or pond. Slinky loops can also be installed at the piers of seas or rivers.

OPEN-WELL OPEN-LOOP SYSTEM

An open-well geothermal system has one or more wells to extract water from the ground. It is important that an adequate quantity of water be available all year round. Large buildings require huge quantities of water. Approximately 3 gallons per minute of water is required for every ton of air conditioning, or every 300 square feet of space to be air-conditioned. Table 10–2 gives the quantities of water required at peak load conditions for air conditioning with standard heat pumps. Standard heat pumps are designed for 3 gallons per minute (gpm) per ton of air conditioning.

The rate of water flow is based on the temperature differential between the supply and return. The water flow rate can be reduced by increasing this temperature differential. However, when the water flow is reduced, the distance between supply and diffusion well is increased. Projects that have limited land should consider this. See Table 10–3.

TABLE 10-2 WATER QUANTITIES FOR OPEN-LOOP SYSTEMS

Area of the Building in Square Feet	Building Type	Peak Cooling Load in Tons	Gallons per Minute (gpm) of Well Water	Approximate Pipe Size in Inches
5,000	Residential	15	45	2
20,000	Small office	65	195	4
50,000	Office building, school, senior living, multifamily residential, industrial	165	500	6
100,000	Large office building, hospital, college campus building, school	330	1,000	8

TABLE 10-3 WATER TEMPERATURE DIFFERENCE AND FLOW RATE

Temperature Difference between Supply and Return	Gallons per Minute (gpm) per Ton of Air Conditioning	Distance between Supply and Diffusion Well
10°F	3	Low
15°F	2	Medium
20°F	1.5	High
30°F	1	Higher

Well water can be pumped directly into the heat pump units. But this is generally avoided, and an isolation heat exchanger is added. The isolation heat exchanger separates the well water loop from the water loop on the building side. The equipment normally used for this is a plate and frame heat exchanger, which transfers the heat between the two loops of water via a stainless steel plate, without physical contact between the two water streams. This isolation heat exchanger prevents any chemicals used for treatment of water from corroding pipes on the heat pump side to contaminate the ground. The groundwater, after it passes through the heat exchanger, is returned to the ground through the diffusion well.

Factors to be considered in the use of open-loop geothermal wells:[11]
- Hydrogeology of the area:
    - ☐ Porosity of the formation
    - ☐ Presence of clay
    - ☐ Hydraulic conductivity

---

[11] "Open loop geothermal loop well systems on Long Island," Paul K. Boyce, P.E., Doreen Fitzsimmons.

- Aquifer knowledge:
  - ☐ Depth of water
  - ☐ Yield
  - ☐ Dissolved solids in water
- Maintenance:
  - ☐ Well maintenance
  - ☐ Piping systems
  - ☐ Heat exchangers

TYPES OF HEAT PUMPS

WATER-TO-AIR HEAT PUMPS

Water-to-air heat pumps are the most common units used with geothermal energy. They are called "water to air" because water from the ground transfers heat to air, which is used to air-condition or heat the building. Several different types of water-to-air heat pumps are available for installation in buildings. The use of heat pump units in buildings has progressed so far that different manufacturers have standardized their product offerings. Table 10–4 gives the types of units available.

Several smaller water-to-air heat pumps can be added to a single water loop. This arrangement is common in a large space of a building with interior and exterior or perimeter zones. This arrangement has the unique advantage of recycling heat from the interior to the exterior or perimeter. Generally, a building floor plan is divided into interior and exterior or perimeter zones. The building perimeter zone is generally a 10-foot to 15-foot depth along the perimeter. The rest of the floor area is the interior zone.

TABLE 10–4 WATER-TO-AIR HEAT PUMP TYPES

Type of Unit	Figure Number	Capacity in Tons	Building Floor Area Covered in Square Feet
Vertical floor	Fig. 10–12	2 to 10	600 to 3,000
Horizontal ceiling	Fig. 10–12 (1)	1 to 5	300 to 1,500
Large commercial	Fig. 10–13	10 to 60	3,000 to 18,000
Floor console	Fig. 10–14	¾ to 1½	225 to 450
Rooftop	Fig. 10–15	3 to 35	900 to 10,500

(1) The ceiling unit is similar to floor except mounted horizontally.

TYPES OF HEAT PUMPS 199

Figure 10–12 Vertical floor-mounted heat pump. *Florida Heat Pump Company*

Figure 10–13 Large commercial heat pump. *Florida Heat Pump Company*

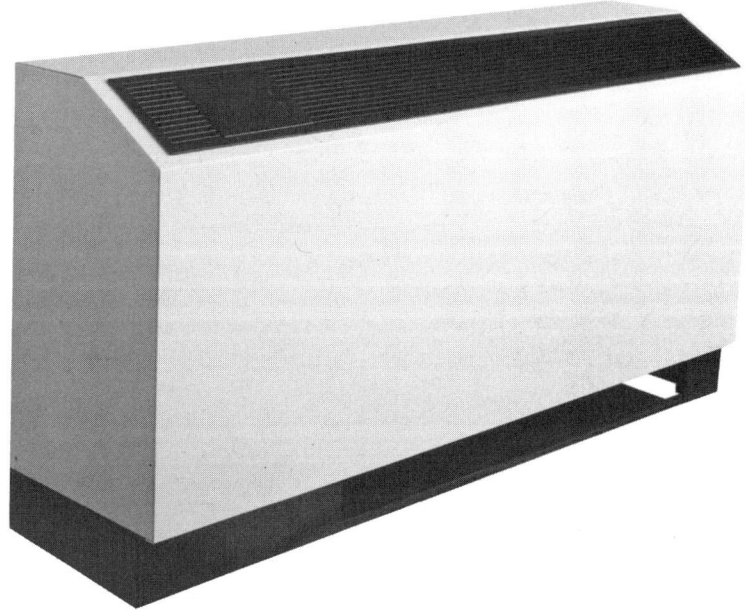

Figure 10–14 Floor-mounted console heat pump. *Florida Heat Pump Company*

Figure 10–15 Rooftop heat pump. *Florida Heat Pump Company*

The perimeter zone heat pumps either heat or cool (cooling in summer and heating in winter). The interior zones always require cooling, both in summer and winter, due to the heat generated from lights, people, and equipment. The interior zones do not have an exterior envelope element where heat is lost. Therefore, even in winter, when exterior zones require heat, the interior zones require cooling. This presents an opportunity to capture heat from the interiors and use it for the exterior, or to recycle heat. Perimeter buildings, which are narrow and long, have only perimeter areas, with a corridor in the middle. Even in perimeter buildings, heat recycling is possible with water source heat pumps—heat from the façade facing the sun can be recycled to the façade that requires heating in winter. The heat pumps provide an opportunity to transfer the interior heat to the perimeter, reducing the use of fossil fuel to generate heat or electrical power to transfer heat from the ground.

The units also provide the flexibility of independently heating and/or cooling individual zones in intermediate seasons. In a conventional chilled water system, the same level of flexibility is obtained only with a four-pipe system, in which two sets of pipes are required (each set with one supply pipe and one return pipe) for cooling and heating. The cost of an additional set of pipes is generally expensive, and in most cases a two-pipe system is chosen. In a two-pipe system, the building is in either heating or cooling mode; it can't do both at the same time. In some designs, electric heating coils are provided to achieve this, but this solution is not environmentally friendly, because air is heated and cooled, wasting energy.

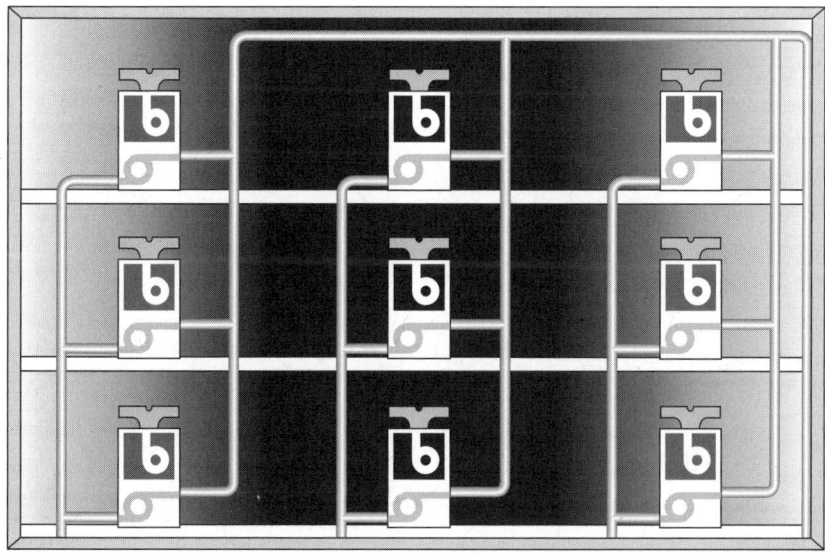

Figure 10–16 Heat transfer from interior to perimeter.

WATER-TO-WATER HEAT PUMPS

Water-to-water heat pumps are less common. Water-to-water heat pumps transfer heat from groundwater to another water loop of building circulating water. Either hot water or chilled water is produced as the building's circulating water loop. The building circulating water is in turn used to condition the building. Conventional fan coils or air handlers can be used similar to the conventional system, however, the source of energy comes from the efficient and renewable geothermal system. Water-to-water heat pumps can also be used for heat recovery from air streams exiting and entering the building. In applications where a large amount of hot or cold contaminated air is exhausted out of the buildings such as in specialty laboratories and fume hoods, the heat from the exhaust can be extracted by a water-to-water heat pump, where water coils in the exhaust air stream acts as source or sink similar to the Earth.

CARBON DIOXIDE HEAT PUMPS

Carbon dioxide heat pumps should not be confused with the greenhouse gas carbon dioxide, which is associated with global warming. In carbon dioxide heat pumps, carbon dioxide is used as the medium of heat transfer. The heat transfer medium, which is usually water, can be replaced with carbon dioxide because carbon dioxide is liquefied at high pressures. The liquefied carbon dioxide evaporates and condenses at the source and sink, without the need for any pump. This is called a heat pipe, whereby

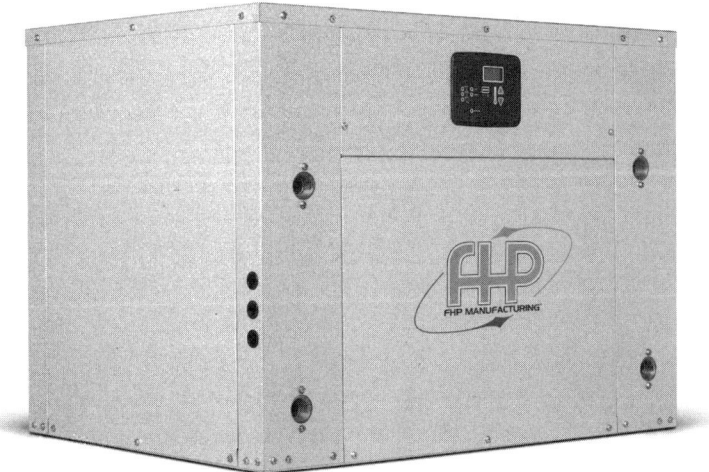

Figure 10–17 Water-to-water geothermal heat pump. *Florida Heat Pump Company.*

evaporation and condensation occur at the two ends of a pipe, which transfers heat from one end to the other. The core benefits of this system are:

- The pumping energy of the water-based system is eliminated.
- Having no moving parts lowers maintenance.
- There is no groundwater contamination.

The disadvantage of this system is that carbon dioxide works as a refrigerant, evaporating and condensing at saturation pressure of 1,200 pounds per square inch (psi). The piping systems have to be built to withstand these high pressures. The building construction industry is generally not used to installing pipes with high pressure of 1,200 psi in buildings. Manufacturers offer products for residential heating and domestic water heating using carbon dioxide. Another industry where carbon dioxide is used as a refrigerant is data center cooling. Carbon dioxide has been used in the pipes in the data center environment, where water is generally avoided near electronic equipment such as server racks. Carbon dioxide acts as a medium, carrying heat from the data center to the outdoors.

CHAPTER 11

Cogeneration

COGENERATION IS ALSO REFERRED TO AS COMBINED HEAT AND power generation (CHP). Cogeneration is the process of simultaneously generating two forms of energy: electrical power and heat. In standard power generation, electrical energy is used, but heat is wasted by exhausting it through the stack. Cogeneration recovers the heat energy that would have been lost and puts it to use. The heat energy can be used in several ways, including space heating in winter, absorption cooling or air conditioning in summer, and domestic hot water use year-round. Another unique feature of cogeneration is that it is localized, or located at the point of use of energy. Cogeneration produces electrical power and heat at the consumer's site, avoiding the purchase of electricity from a utility company. Electrical generation with fossil fuel is normally done via gas boilers, which burn to produce heat; the heat produces steam to drive the turbines, which in turn power the generators, producing electricity. Another form of electrical power generation with fossil fuels uses direct-fired equipment such as gas turbines and internal combustion engines. Fossil fuel is burned directly in the engine or turbine, which powers the generators, producing electricity. In both turbines and internal combustion engines, heat energy is converted into electrical energy. The process of conversion of heat energy to electrical energy is only 30 to 40 percent efficient; 60 to 70 percent of the heat is released to the atmosphere via a flue stack or cooling tower. Of all the energy produced in the United States, 63.9 percent is lost in the conversion process;[1] or, two-thirds of the fuel used to generate power is lost as heat.

---

[1] U.S. Energy Information Administration, Annual Energy Review, 2007.

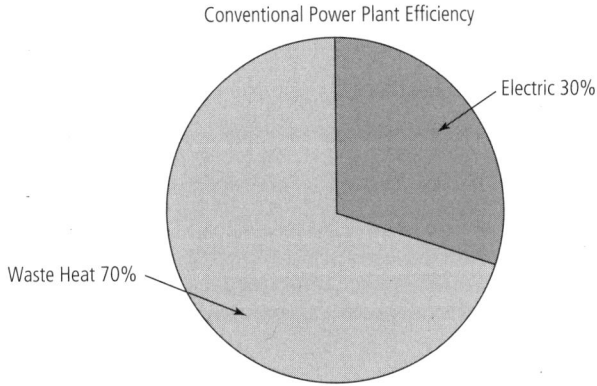

Figure 11–1 Conventional power plant efficiency. *Asif Syed*

The location of the plant where power is generated is important, if the waste heat has to be captured and used. Most large industrial power plants are located far from population centers or industrial centers, in areas where waste heat cannot be put to use. The locations of power plants are generally determined by the availability and transportation of fuel. In some instances, power plants located near population centers and industrial centers do capture and use the waste heat. In New York City, waste heat from a power plant is used to generate high-pressure steam, which is distributed via a piping network to buildings in Manhattan. This steam is used for heating and cooling the buildings: In winter, steam is used for heating, after its pressure is adjusted in pressure-reducing valves; it can generate hot water or be used directly in radiators. In summer, high-pressure steam is used in steam turbine–driven chillers or absorption chillers. In the Middle East, cities such as Dubai, Kuwait, and Jeddah use the waste heat from power plants to produce desalinated water. However, in most power plants in the United States and in other parts of the world, waste heat is not recovered. The average site-to-source energy ratio is 3.6—that is, about 3.6 units of fossil fuel energy are consumed for every 1 unit of electrical energy delivered to the site or building.

An average cogeneration plant can deliver 60 to 70 percent efficiency, double that of conventional power generation, and a well-designed plant can achieve efficiency of 90 percent. The waste heat is utilized in the building to heat and cool. In summer, the heat is used for air conditioning; in winter, heat is used for building heating; and heat can be used year-round for domestic hot water needs in the building. Absorption chillers use waste heat from the cogeneration plant to produce chilled water. Absorption refrigeration uses heat to provide the energy required for the cooling process. Absorption chiller technology is a proven method of air conditioning commonly used in buildings. Large absorption chillers, independent of cogeneration, have been in use in buildings since the 1960s.

Cogeneration economics are generally profitable and have a good rate of return or payback on investment. In addition, most governments realize the benefits of

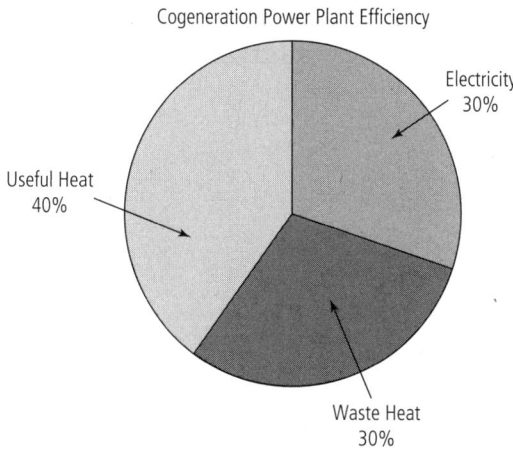

Figure 11–2 Cogeneration power plant efficiency. *Asif Syed*

cogeneration and offer incentives to encourage installation of plants. These incentives are normally rebates on the installation costs. Most utility companies also realize that cogeneration plants help control their investment costs for the infrastructure of power generation and distribution facilities. The expense of building new power plants and power distribution networks—and the difficulty of obtaining the necessary permits—can be avoided with cogeneration.

OTHER APPLICATIONS OF COGENERATION

Landfills: Cogeneration can be used in landfills. Using the waste gases generated from landfills, electricity is generally exported to the utility grid, and heat or hot water is exported to the nearby sites. The cost of fossil fuel is completely eliminated, because these waste gases are a free by-product of landfills. The only cost associated with the fuel is the investment for the gas-capturing system.

Sewage treatment plants: Waste gases generated from sewage treatment plants are processed and used in cogeneration plants to generate electricity and heat. Electricity is generally used in the plant and also exported to the utility grid, and heat or hot water can be used in the sewage treatment process, for digesters or sludge drying.

Greenhouses: Greenhouses need heat in winter and carbon dioxide for photosynthesis. Normally, fossil fuel is used to generate heat. With cogeneration, heat is used in the greenhouse, and the carbon dioxide from the exhaust gases is captured and pumped into the greenhouse to accelerate photosynthesis. Electricity generated is exported to the utility grid.

Cattle farms: Manure generated in cattle farms is a good source of methane or biogas. The biogas is used in cogeneration plants to produce electricity and heat, which are generally used on the farm, and excess electricity is exported into the utility grid. Cattle farm cogeneration recycles the manure as fertilizer.

BENEFITS OF COGENERATION

The benefits of cogeneration are:

1. Higher efficiency
2. Lower operating cost
3. Clean and environmentally friendly power
4. Reduction of power distribution infrastructure
5. Reduction of transmission losses
6. Reduction of carbon footprint

These benefits solve several challenges facing the power industry. Greenhouse gases, especially carbon dioxide, can be reduced to half with cogeneration. By improving efficiency, cogeneration reduces the cost of operation, making power generation more cost effective, and making businesses competitive. Utility companies that need to improve their infrastructure require huge investments, which can be avoided totally with cogeneration. The risks involved with speculative planning of large-scale infrastructure development can be avoided because cogeneration is localized—it produces power where required and used.

Higher efficiency: In buildings that use heat, air conditioning, and electricity, the operating efficiency of the system can be doubled. The operating efficiency of a boiler plant that purchases electrical power is approximately 48 percent when electricity is purchased from a utility company and heat is generated in the boiler system. To produce 40 units of electricity, about 110 units of fuel are consumed. The losses are from the stack gases that are vented to the atmosphere and the distribution losses from the power plant to the building site. And to produce 30 units of heat, about 35 units of fossil fuel are consumed. The losses are due to boiler operating inefficiencies. In the conventional approach, to purchase electricity and generate heat input of 145 units is required for an output of 70 units. Thus, the operating efficiency of the conventional system is 48 percent. A combined heat and power (CHP), or cogeneration, system will produce the same 40 units of electricity and 30 units of heat with only 100 units of fossil fuel. The operating efficiency of the cogeneration system is 70 percent. All the numbers are rounded off for simplicity, but they are within a plus or minus range of 2 to 5 percent.

Lower operating cost: Historically, the cost of natural gas has been lower than that of purchased electric power, per unit of energy. The average price of gas per 1,000 cubic feet is $6.73. The cost of gas per therm, the commonly used gas unit,

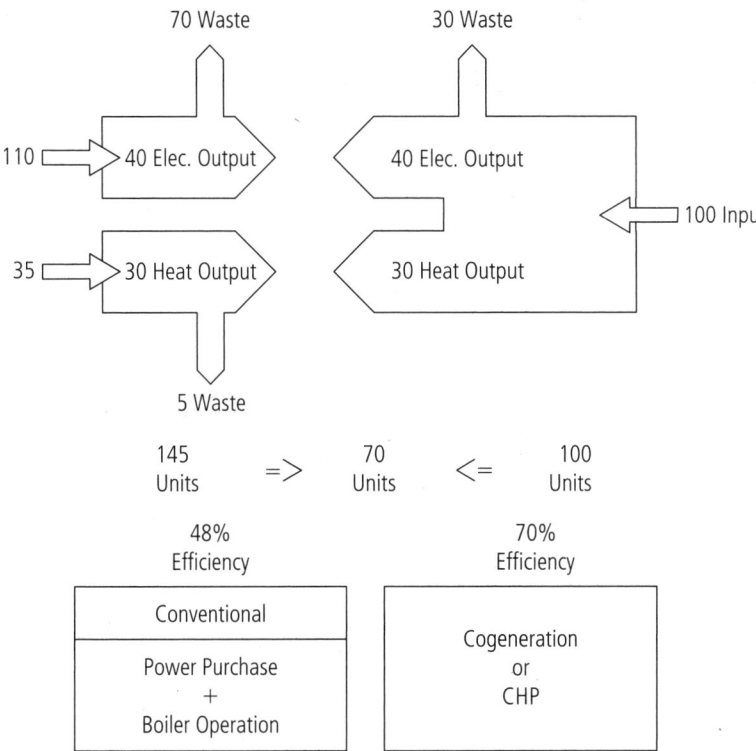

Figure 11–3 Operational efficiency of cogeneration, or combined heat and power, system. *Asif Syed*

is $0.673. The cost of producing electricity with natural gas can be calculated as follows:

$$\$.673 = 100 \text{ cubic feet of gas}$$
$$= 1 \text{ therm}$$
$$= 100{,}000 \text{ Btu/hr}$$
$$= 29.41 \text{ kW } (3{,}400 \text{ Btu/hr} = 1 \text{ kWh})$$
$$= 11.76 \text{ kWh of electricity produced (40\% efficient generation)}$$

Free heat = 30 cubic feet of gas (30% heat capture)
$$= \$0.202$$

($ spent on gas − $ of useful heat captured = $ to produce electricity)

($0.673 − $0.202 = $0.471)

$0.471 = 11.76 KW of electricity

1 kWh = $ 0.04 (0.471/11.76)

The cost of gas used here is the average price from 2005 to 2010. It is in the range of the current prices for 2011. The average cost of electricity in the United States is $0.10 per kilowatt hour, which is double the price of producing electricity with natural gas by cogeneration. In some states, the price of electricity can be as high as $0.13 to $0.29 per kilowatt hour.[2] Every site or building location has different electric and gas rate structures, and a study is required to establish the cost of cogeneration and the cost of electric purchase for a particular location. The calculations above are only for rule of thumb demonstrative purposes. An actual cogeneration analysis involves hour by hour calculation between electric and heat demand, cost of electricity and gas (commodity and distribution), any surcharges paid for stand-by power from utility, and any down time of cogeneration plant. Those calculations will provide an annual thermal and electrical efficiency, whereas the rule of thumb is for an instantaneous hour.

Clean and environmentally friendly: The two major environmental benefits of cogeneration technologies in buildings are:

1. Lower levels of greenhouse gases or carbon emissions
2. Lower levels of air pollution

Reduction of greenhouse gases (GHG) and carbon emissions is due to the capturing of the heat of exhaust gases. The exhaust gases in a conventional utility power plant that feeds the power grid are usually released into the atmosphere. Capturing exhaust gases and using them to heat or cool the buildings eliminates the burning of fossil fuel that would have been required in a conventional system.

TABLE 11-1 NATURAL GAS INDUSTRIAL PRICE $ PER 1,000 CUBIC FEET[3]

Year	Jan	Feb	Mar	Apr	May	Jun	Jul	Aug	Sep	Oct	Nov	Dec
2001	8.84	7.21	6.3	6.08	5.46	4.8	4.1	3.99	3.5	3.18	3.88	3.69
2002	4.05	3.7	3.78	3.64	4.07	3.9	3.8	3.62	3.89	4.18	4.72	4.92
2003	5.65	6.4	8.27	5.96	5.78	6.6	5.7	5.28	5.32	4.93	5.19	5.9
2004	6.72	6.52	5.97	6.06	6.34	6.8	6.4	6.36	5.68	6.03	7.64	7.54
2005	7.06	7.15	7.12	7.71	7.19	6.9	7.4	7.98	10.2	12.1	12.1	11.2
2006	10.9	9.38	8.24	7.93	7.63	6.9	6.8	7.36	7.21	5.62	7.74	8.23
2007	7.36	8.25	8.42	8.14	8.11	7.9	7.5	6.72	6.28	7.06	7.87	8.18
2008	8.33	9	9.64	10.1	11.4	12	13	10.1	9.13	8.11	7.36	7.89
2009	7.43	6.37	5.65	5.03	4.35	4.5	4.6	4.31	3.81	4.8	5.37	5.97

---

[2] U.S. Energy Information Administration, State Electricity Profiles, 2008.
[3] U.S. Energy Information Administration, Independent Statistics & Analysis.

The United States Department of Energy estimates[4] that if 20 percent of electricity generation capacity comes from CHP by 2030, there will be a 12 percent annual greenhouse gas CO_2 reduction of 848 million metric tons (MMT)—the equivalent of taking 154 million cars off the road. The reduction in carbon dioxide emissions equates to about 60 percent of the projected growth in emissions.

The local environmental benefit of cogeneration is cleaner air quality. Cogeneration plants located within a building or campus of buildings burn fossil fuel to produce electricity. The products of combustion are released into the atmosphere. The combustion products include sulfur dioxide (SO_2), also known as SOx, and various nitrous oxides, also known as NOx. Both NOx and SOx are environmentally harmful pollutants. The release of NOx and SOx are regulated by state and federal agencies. Cogeneration plants have to comply with Environmental Protection Agency (EPA) regulations for National Ambient Air Quality Standards (NAAQS). Control of NOx and SOx is accomplished with scrubbers, which are devices similar to catalytic convertors in automobiles. Most of the existing power plants that supply electric power to utility grids do not have the same level of control of NOx and SOx, due to preexisting or grandfathered conditions. Therefore, production of the same power by a new cogeneration plant creates lower air emissions or pollution. The technology to control NOx and SOx emissions adds cost to the cogeneration plant. Cogeneration can become more competitive if NAAQS controls on NOx and SOx are required for all power utility plants, both new and existing.

COGENERATION TECHNOLOGIES

In all cogeneration systems, electricity is produced in a generator, which is driven by rotating equipment also commonly referred to as the prime mover. In different technologies the prime mover is different, but all of them have the primary goal of rotating an electrical generator to produce electricity. In all technologies, fossil fuels are used. Natural gas is the preferred fuel due to its clean-burning nature. However, other fossil fuels, such as diesel, propane, gasoline, alternate fuels, and coal, can be used. Waste gases from other processes such as waste water treatment, landfills, dairy industry, and so on can also be used for cogeneration. The most common technologies available are:

1. Reciprocating engines
2. Gas turbines

---

[4] "Combined Heat and Power: Effective Energy Solutions for a Sustainable Future," report prepared by Oak Ridge National Laboratory for the U.S. Dept. of Energy, December 1, 2008.

3. Micro turbines
4. Fuel cells
5. Steam turbines

The efficiencies and applications of these technologies vary depending primarily on the size of the system. Micro turbines are smaller systems suitable for smaller buildings. Reciprocating engines and gas turbines are for medium to large facilities. Gas and steam turbines are generally used in larger systems such as district power and cooling plants. Steam turbine plants require a boiler plant to produce steam—thus increasing the cost. In places where excess steam is available, steam turbines are an excellent choice. Steam turbines are one of the oldest technologies for power generation in the industrial and electrical power utility sectors. Steam turbines are rarely seen in commercial buildings.

RECIPROCATING ENGINE COGENERATION

Reciprocating engines are the most common form of cogeneration system in buildings. Reciprocating engine technology is familiar to the building industry from life-safety and emergency standby generators. Their familiarity in the building sector means that specialized trained personnel are not needed to operate and maintain them. Natural gas is the preferred fuel for a reciprocating engine cogeneration system; however, diesel or dual-fuel diesel and natural gas can be used. The average efficiency of the diesel engine generators is as follows:

TABLE 11-2 CAPACITY RANGE; ELECTRICAL POWER AND TOTAL (HEAT AND ELECTRICAL POWER) EFFICIENCIES; AND COST OF COGENERATION TECHNOLOGIES[5]

Cogen Technology	Capacity in MW	Electrical Power Efficiency	Total Efficiency (Power + Heat)	Cost $/kWh
Reciprocating engine	0.25–5	22–40%	70–80%	$1,100–$2,200
Gas turbine	1–15	22–36%	70–75%	$970–$1,300 (5–40 MW)
Micro-turbine	0.4–0.7	18–27%	65–75%	$2,400–$3,000
Fuel cell	1–2	30–60%	55–80 %	$5,000–$6,500

Note: The capacity of gas turbines is much higher. The 15 MW limit is generally for commercial buildings or district plants. Gas turbine capacity can be up to 250 MW.

---

[5] Catalog of CHP technologies, United States Environmental Protection Agency, Combined heat and Power partnership, December 2008

1. Electrical efficiency = 40% (electricity produced)
2. Heat capture efficiency = 30% (jacket water = 15%, and exhaust gases = 15%)
3. Exhaust gases = 30% (heat vented to atmosphere)

Waste heat recovered from reciprocating engines is from two sources: jacket water and exhaust stack gases. The jacket water is the liquid circulating medium that cools the engine block. The heat recovered from the jacket water is below the boiling point of water, or less than 212°F. It is also referred to as low-grade heat. Exhaust stack gases are products of combustion that are exhausted from the engine at temperatures in the range of 900 to 1,000°F. The high temperature of the exhaust gases makes this high-grade heat suitable to produce steam. The quantity of heat recovered from the jacket water is 50 percent, and the quantity of heat recovered from exhaust gases is approximately 50 percent.

The most common fuel used for cogeneration with reciprocating engines is natural gas. However, diesel is a common fuel for reciprocating engines for standby power and life-safety power. Life-safety power is the power required to operate the life-safety systems in the buildings necessary to safely evacuate the occupants. Some systems associated with life safety are fire pumps for sprinklers, egress lighting, smoke exhaust, and so on. Standby power is used when there is a blackout or brownout, to allow the business to continue to operate. The extent of standby power varies according to the business's needs. Most commercial, institutional, and multifamily large residential buildings have a reciprocating engine for life-safety power. Most standby and life-safety generators use diesel as the fuel for the reciprocating engines. Diesel is easy to store in the building. The quantity of storage for life-safety applications is limited to 6 hours. In standby applications, diesel generators serve as backups to continue business operations when there is a wider power outage. The storage for standby applications can vary depending on the business's needs, but is generally in the range of 24 to 48 hours.

Availability of natural gas at sites is relatively common. There is a very good and reliable natural gas infrastructure. According to the U.S. Department of Energy (DOE), "Reliable, economic, and flexible energy transportation infrastructure is essential to national security and economic prosperity." DOE goals are to enhance gas delivery reliability, reduce environmental impact by increasing the capacity of existing infrastructure, address gas and electric power interdependencies and infrastructure, and develop new technology for intelligent gas delivery systems.[6] With the availability and reliability of natural gas, increasing bi-fuel and dual-fuel engine technologies have

---

[6] United States Department of Energy (DOE) online, Oil and Natural Gas Technologies: Transmission, Distribution & Storage.

developed. Bi-fuels use natural gas or diesel (not both at the same time). Dual-fuels use natural and diesel at the same time (a mixture of both). Existing single-fuel diesel engines can be converted into bi-fuel or dual-fuel engines with conversion kits.

The synergies of:

1. Availability and reliability of natural gas,
2. Existing or required engines in buildings,
3. Dual-fuel capability of engines, and
4. Higher cost of power at peak periods

provide a perfect opportunity for cogeneration and dispatchable standby power generation.

Dispatchable standby generation (DSG) is the generation of power at the site using existing standby or life-safety generators during peak power demand periods. DSG does not provide the benefit of improving the efficiency of the combustion process, as in cogeneration. However, it has both environmental and economic benefits. The peak electrical demand occurs when most consumers are using electricity. The peak demand usually happens in summer months, when air conditioning is added to normal power usage. During these peak periods, the cost of electric power goes up, due to supply and demand, and it can be several times the nonpeak cost. DSG provides an opportunity to operate the standby and life-safety reciprocating engines in buildings. DSG benefits both building owners and utility companies. The utility companies call the building operators, usually 24 hours ahead, and advise them to operate the generators for 4 to 5 hours the next day. For the entire year, the total peak demand hours are usually about 200 hours. To operate standby and life-safety generators for DSG, additional equipment and controls have to be installed. The additional equipment required includes automatic transfer switches, paralleling breakers, synchronizing and power quality controls, and safety relays for interconnection with the utility power grid. Most utilities have a DSG program, as it benefits them by lowering

TABLE 11-3 BENEFITS OF DSG

Building Owners	Utility Companies
Reduces the cost of power at peak demand hours	Improves reliability of power
Exercises idle equipment and improves reliability	Reduces the cost of infrastructure for power plants and distribution
Cost shared or paid by utility company	Avoid operating peak capacity plants
Reduces greenhouse gas (GHG) emissions	Reduces greenhouse gas (GHG) emissions

their investment in infrastructure. The cost of additional equipment, controls, and designs is cost-shared or borne by the utility companies. The benefits to building owners and power utility companies are:

Cogeneration combines the benefits of dispatchable standby generation and the heat recovery from combustion. Cogeneration offers all the benefits of DSG and, beyond that, increases overall efficiency from 30 to 40 percent to 70 to 80 percent, benefiting the environment and economics. The increase in efficiency reduces the greenhouse gases and avoids purchase of energy (by recovering it), reducing operating costs. Cogeneration requires integration with other building systems. Space planning is required to locate the generators and the discharge of exhaust. Mechanical and electrical systems such as air conditioning, domestic water heating, and space heating must be coordinated and integrated with cogeneration systems. The heat recovered from the cogeneration system is used for heating domestic hot water, heating the building space, and air conditioning (using absorption chillers). Cogeneration can be introduced in existing buildings as well as new buildings. Prior to installing a cogeneration system, a site-specific cogeneration feasibility study has to be preformed.

Facilities that have a high percentage of standby power are:
- Telecommunications buildings
- Financial institutions
- Data centers
- Broadcast facilities
- Hospitals
- Hospitality industry buildings
- Airports

Most of the listed facilities are also in use 24/7, and have much or all of their power backed up with standby emergency reciprocating engine generators. The standby generators are important to the continuity of the operation of the facility in a blackout or brownout, or a localized power failure event. The investment in the standby generators is warranted by the business's needs. However, it is the most underutilized asset. The majority of the outages experienced by customers in the United States are due to localized events, rather than large events that affect the bulk power systems. The infrastructure of both power generation and distribution in the United States generally has the capacity for the peak demand. The underutilized asset can be easily put to use for cogeneration and or dispatchable standby generation.

Continuous, Prime, and Standby Generators

The main difference between standby and prime and continuous reciprocating engines is the revolutions per minute (rpm) at which they operate. Standby generators

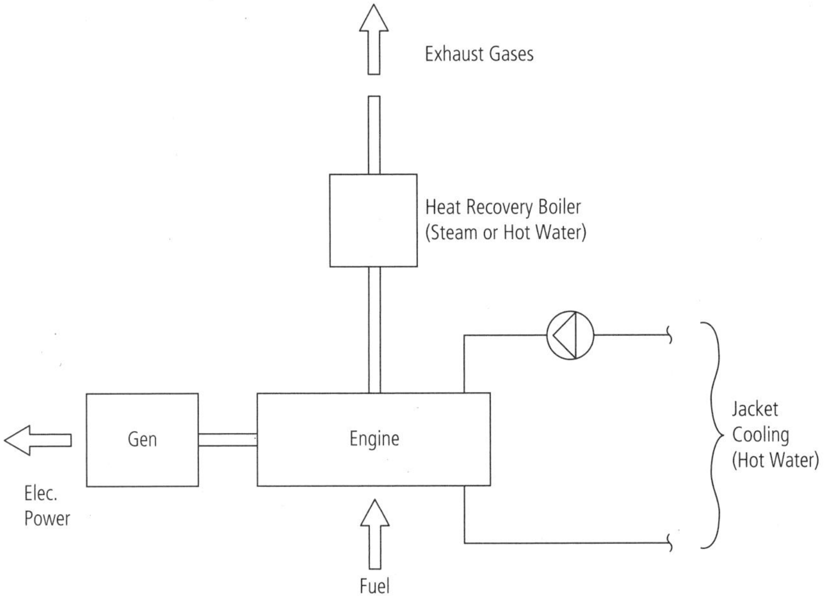

Figure 11–4 Reciprocating engine cogeneration. *Asif Syed*

are meant to operate from 50 to 200 hours per year. Most of these hours consist of exercising the generator periodically to ensure startup in an emergency. Other hours include downtime in electrical power. Most standby generator engines operate at 1,800 rpm. Prime and continuous generator engines operate for longer periods of time. Prime power engines operate 8 to 12 hours per day. Continuous power engines operate 24 hours a day and 7 days a week. Both prime and continuous power generator engines operate at a lower 800 to 1,100 rpm. The lower rpm reduces the wear on the engine, increases life, and reduces maintenance cost; but lower rpm generators cost more.

GAS TURBINE COGENERATION

Gas turbine is a proven technology for producing electrical power. It is widely used in industrial power generation and by utility companies. The gas turbines used in buildings are smaller, in the 1 to 10 megawatt (MW) range. The advantage of gas turbines over reciprocating engines is their high-grade heat. Turbines produce a single source of heat from the exhaust gases, and its temperature is well above the boiling point of water to produce steam. Recovery of high-grade heat via production of steam eliminates any hot water systems such as pumps. Steam can be easily transported and connected into existing steam distribution systems. Steam can also be used in steam turbine generators to generate electricity. The most common fuel used for gas

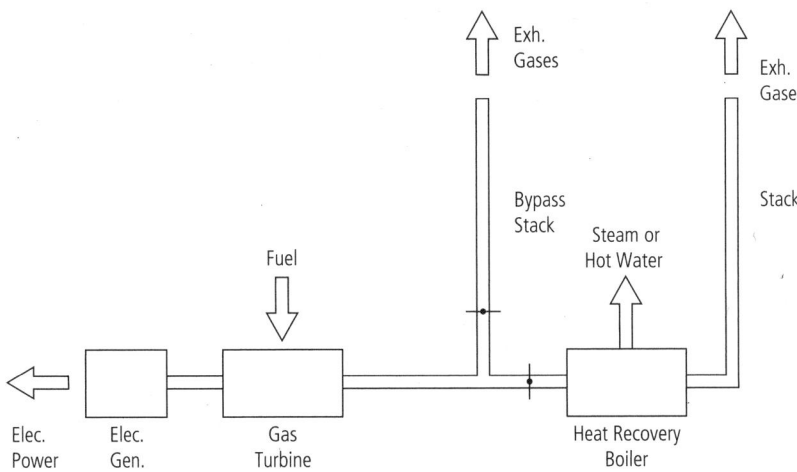

Figure 11–5 Gas turbine cogeneration. *Asif Syed*

turbines is natural gas. However, other forms of fuel such as diesel and landfill methane (CH_4) can be used.

Typical applications for gas turbines are a district cooling plant and a heating plant connected to several buildings, such as a college or university complex of several buildings. Princeton University has a 15-MW cogeneration plant that feeds the campus with electrical power and heat.[7] Gas turbines in the 5 to 10-MW range are also seen in commercial buildings. Bank of America Tower, a 54-floor high-rise commercial building at One Bryant Park in the heart of New York City, has a 5.0-MW gas turbine cogeneration plant.[8] The new University Medical Center of Princeton at Plainsboro will use a 4.6-MW gas turbine cogeneration plant.[9] Gas turbine cogeneration plants can be located in self-contained buildings, as at Princeton University. However, gas turbine cogeneration plants can be integrated into buildings, as in the Bank of America Tower. The cost of gas turbine installation is lower at large turbine sizes. Smaller gas turbines of 1 MW have an installed cost of $3,300 per kilowatt, whereas larger gas turbines of 5 to 10 MW have an installed cost of $1,300 per kilowatt.[10] This makes gas turbines above 5 MW more attractive and economically viable. In most gas turbine applications, electrical power is parallel and synchronized with the utility; the steam is

---

[7] Princeton University website, Facilities, Princeton Energy Plant. www.princeton.edu/facilities/info/major_projects/energy_plant/.
[8] Bank of America website, "Bank of America Tower at One Bryant Park: Reaching New Heights in New York," June 11, 2009.
[9] "'Topping Out' Ceremony Marks Milestone in Construction of UMCPP," Princeton Health Care System News, December 3, 2009 (www.princetonhcs.org/Default.aspx?p=4089&d=3743).
[10] "Technology Characterization: Gas Turbines," prepared by Energy and Environmental Analysis for the Environmental Protection Agency, December 2008.

Figure 11–6 Cogeneration plant at Princeton University. *Asif Syed*

Figure 11–7 Cogeneration plant in a commercial building: Bank of America Tower. *Asif Syed*

distributed to the campus heating grid to be used for heating in winter, domestic hot water use, and air conditioning in summer using absorption chillers.

MICRO TURBINE COGENERATION

As their name suggests, micro turbines are small gas turbines. Micro turbines are prepackaged with all components in an assembled box. Interconnection design between turbines and the heat recovery boiler, as in gas turbines, is not required. All components such as gas turbines, gas compressors, heat recovery devices, and electrical components are packaged in a 6-foot by 4-foot by 3-foot (approximate indicative size) box. Micro turbine sizes range from 30 kW to 250 kW. Micro turbines are modular, which allows multiple units to be combined to create larger capacity. The modularity allows for part loading or shutdown of each unit, while the other unit operates at full capacity. Micro turbines produce heat that can be used for domestic hot water heating, building heating, or air conditioning with absorption chillers. The most common fuel used in micro turbines is natural gas; however, other liquid fuels such as diesel can be used. Micro turbines also use landfill-recovered waste gases.

The most common applications for micro turbines are multifamily residential buildings, hotels, small hospitals, and college dormitories. These buildings have a 24/7 demand for heat for domestic hot water. The micro turbines are sized such that all the heat and electricity is utilized. Heat discharge to the environment due to lack of demand reduces the annual efficiency of the turbine. Usually hot water storage tanks can be used to store heat during the low-demand times, to be used at peak demand. In a hotel or college dormitory, the early morning hot water demand is much higher than the demand during late-night hours. Proper sizing of storage tanks and micro turbines increases the annual efficiency and accelerates the payback or return on investment (ROI). Micro turbines that operate 24/7 and 365 days a year have the fastest payback or ROI. Micro turbines can be used for 10 to 12 hours a day in office buildings where there is a demand for heat, or heat can be exported to a neighboring building where there is a demand. Buildings that operate turbines for fewer hours have a longer payback.

FUEL CELL COGENERATION

Fuel cells produce electricity without combustion, unlike all other technologies, which involve combustion of fossil fuel. Fuel cells produce electricity from a hydrocarbon fuel, in an electrochemical process using an electrolyte (or polymer electrolyte membrane). The hydrocarbon fuel is processed in a fuel cell processor into pure hydrogen. This hydrogen atom is changed into positive hydrogen protons and negative hydrogen electrons. The electrolyte prevents electrons from flowing through, allowing protons to pass through. The negative electrons follow the route through a conductor

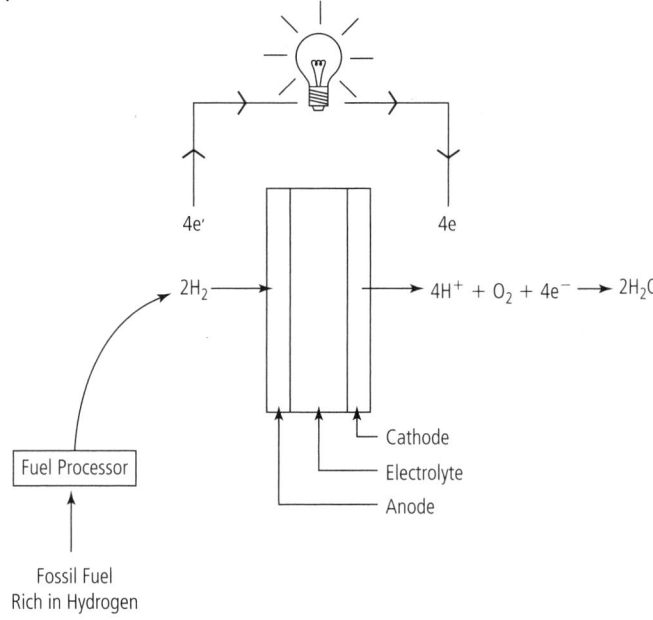

Figure 11–8 Fuel cell operation. *Asif Syed*

(electrical wire), thereby producing electricity. At the upstream of the electrolyte is the cathode, and at the downstream is the anode. At the cathode, the electrons, which have passed through the conductor (light bulb), combine with the hydrogen protons and oxygen (from the atmosphere) to form water.

Hydrogen fuel cells are the most promising technology to replace conventional power generation. This technology is flexible enough to be adapted to both large-scale uses, such as industrial plants, and small-scale applications such as automobiles. Commercial buildings can easily adopt it, because fossil fuel such as natural gas is readily available on-site, and the waste heat can be used for air conditioning, building heating, and domestic water. Hydrogen, the primary fuel, can be processed using renewable energy such as solar power and stored to be used during the night.

At present, the cost of fuel cells is higher than that of conventional power generation with fossil fuel such as coal and oil. However, as the technology becomes more common and production of units increases, the cost will come down. Initially, the cost of fuel cell technology—when it was used in space programs in the 1970s—was $600,000 per kilowatt; however, the present cost is $4,500 per kilowatt.

The Solid State Energy Conversion Alliance (SECA), which is made up of private sector and government research and development groups, is working to reduce the cost of fuel cells. The goal is to bring down the cost to $400 per kilowatt, so that it is competitive with other technologies.

TABLE 11-4 COMPARATIVE COST OF FUEL CELL[11] SYSTEMS

Fuel cells	$4,500/kW
Diesel generators	$800–1500/kW
Natural gas turbines	$4,000/kW

MICRO-COGENERATION OR COMBINED HEAT AND POWER (MICRO-CHP)

Micro-cogeneration, or micro-CHP, is generally defined as power capacity that is suitable for one single-family home. CHP systems less than 15 kilowatts fall into this category. Micro-CHP must not be confused with micro turbines. Micro turbines are smaller versions of direct gas-fired turbines, whereas micro-cogeneration, or micro-CHP, is a complete heat and electrical energy generation process. Micro-CHP systems are heat driven, and electricity is a by-product that can be exported to the grid or stored in a battery. In large industrial systems, the primary driver of the plant is electrical power generation, and heat is a by-product. The plant operates to meet the electrical power demand, and heat is used as required or released into the environment. The most common fuel for micro-cogeneration is piped natural gas, readily available in most residential buildings.

Micro-CHP is not popular yet, but there are several manufacturers offering standard products. The National Institute of Standards and Technology[12] is developing standards for measurement of the efficiency for micro-cogeneration. Presently there are no standards or testing methods to uniformly measure its efficiency and benefits. Testing and rating procedures are being developed by NIST by performance measurements of micro-CHP in 2012. ASHRAE is developing a standard for a testing and rating method for micro-CHP; NIST data will be used to validate this standard. Standards are expected in the future; then consumers who are single-family homeowners will be able to evaluate alternative technologies, products, and manufacturers.

Sizing of micro-cogeneration in homes is complex because several variables are involved. However, it is important to optimize the size to maximize the benefit. The heating load is the primary driving factor in sizing. Heat storage capacity is important, because heat may not be used immediately. The heat load varies, depending on the

---

[11] United States Department of Energy, Fuel Cells Research and Development.
[12] Development of a Rating Methodology for Micro-Cogeneration Technologies Project, The National Institute of Standards and Technology (NIST).

climate zone and the size and construction of the house. The electricity cost structure and the cost of gas or cogeneration fuel are important in the economics and sizing of the micro-cogeneration installation. The utility company's policy on net metering, or the price at which electricity is purchased back, also plays into the economics.

COGENERATION FEASIBILITY STUDY

A cogeneration feasibility study is the first step toward a cogeneration installation. This study should include:

1. Cost of producing electricity
2. Cost of purchasing electricity from the utility company
3. Capital cost of installation
4. Operational and maintenance cost
5. Environmental impact (local noise and emissions)
6. Carbon footprint reduction
7. Hour-by-hour study for 365 days
8. Payback on investment
9. Incentives offered by local and federal governments and utility companies
10. Investment tax credits

CHAPTER 12

Data Center Sustainability

DATA CENTERS HAVE BECOME AN IMPORTANT PART OF OUR LIVES. We use electronic data more frequently than ever before, and the potential uses for this technology are even greater. There is no need to explain how our lives revolve around the Internet and access to data for travel, news, entrainment, business, communications, commerce, education, and so forth. Almost every aspect of our lives—economic, political, or private—relies on data today, and this reliance on data will only grow. Behind all this, there are large buildings that house the data centers. These data centers use enormous amounts of energy to run the servers and to provide the environment necessary for them to function reliably and continuously, without downtime. A standard building has a heating load of 5 watts per square foot, whereas data center loads can be as high as 150 to 200 watts per square foot, making data centers up to forty times more energy-intensive than standard buildings. Creating an environment in which servers can operate requires energy for air conditioning equipment. Creating a reliable power source also takes energy; in addition to the servers, the uninterruptible power supply (UPS) requires power and has losses. Most air conditioning systems used for data centers have very low energy efficiency; their designs evolved around variables other than energy, and energy efficiency was not a priority. The HVAC systems used for cooling a process plant or an office building, school, or hospital are more efficient than those used for data centers. It takes far less energy to remove the same amount of heat from an office building than from a data center. As data centers are taking a center stage in our lives, there has been an awakening to and awareness of their energy consumption and lack of energy efficiency. This chapter discusses the major areas where energy efficiency can be increased.

Companies are gathering data on energy use because part of their corporate environmental responsibility is to reduce energy costs. Data centers are a significant part of an organization's energy consumption, and reducing their energy use provides an opportunity to not only reduce greenhouse gases but also lower costs.

HISTORY OF DATA CENTERS

The present data centers were a product of the dot-com boom of the late 1990s. The time line of the dot-com bubble started in 1994 with the founding of Amazon.com, the largest U.S. online retailer, and it burst in March 2004, when NASDAQ collapsed from its peak of 5,000. During this period, there was a rapid expansion of computing centers or data centers. The mere six years of the bubble was not long enough for the construction industry to think through a perfect solution for the particular needs of data centers. Work had to be done fast, so existing solutions had to be used to fill the ever-growing demand for data centers. There was no time to focus on the inefficiencies of energy consumption. The origin of data centers[1] goes back to the early computers of the 1970s. Because the early computers were large and required a controlled environment, they were located in isolated rooms. The boom of microcomputers in the 1980s led to the client servers of the 1990s, which found a home in the older computer rooms designed for the original mainframes—a totally different computer. The environment of the computer rooms became the standard for server rooms, but microcomputer servers do not need the same highly controlled environment. This resulted in cooling inefficiencies that multiplied with the exponential growth of data centers during the dot-com bubble. This has led to data centers with a power utility efficiency of 2.5 (average);[2] that is, as measured by the building electric utility meter, it takes 2.5 times more energy to run the data center than the energy used for computing by the computers.

An EPA study[3] on data centers reports the following reasons for the growth of data centers:
- Increased use of electronic financial transactions
- Online banking and electronic trading
- Internet communication and entertainment
- Electronic medical records for health-care
- Global economic growth of commerce and services

---

[1] Amy Nutt, "History of the Data Centre," www.morphosppc.com/article/history-of-the-data-centre.
[2] The Uptime Institute, http://searchdatacenter.techtarget.com/definition/power-usage-effectiveness-PUE.
[3] Report to Congress on Server and Data Center Energy Efficiency, Public Law 109-431, U.S. Environmental Protection Agency ENERGY STAR Program, August 2, 2007.

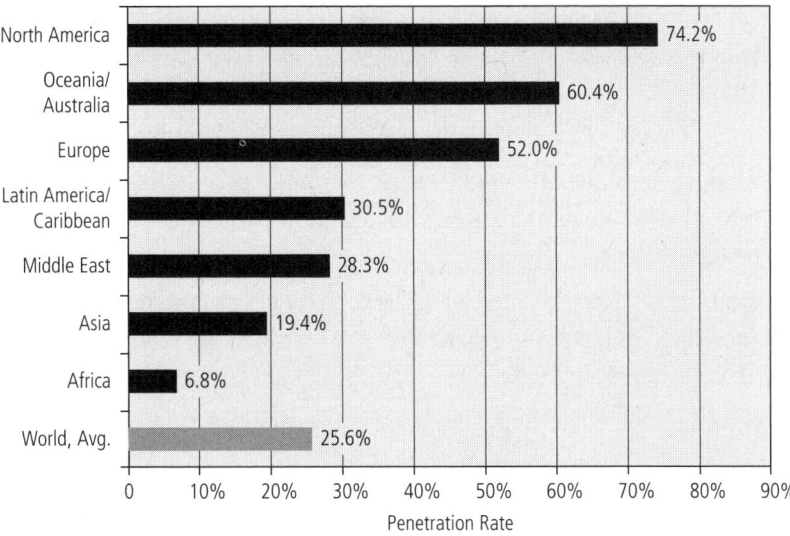

Figure 12–1 World Internet penetration rates by geographic regions. *Internet World Stats, www.internetworldstats.com/stats.htm. Penetration Rates are based on a world population of 6,767,805,208 and 1,733,993,741 estimated Internet users for September 30, 2009. Copyright 2009, Miniwatts Marketing Group.*

- Satellite GPS navigation
- Shipment tracking
- Growth in social networking
- Government information published on the Internet
- Government regulations requiring digital records retention
- High-performance scientific computing
- Growth in Voice over Internet Protocol
- Future growth as more people get digital; only 27 percent of the world population is connected on the Internet[4]

2011: TOP TEN TRENDS IN DATA CENTERS

The top ten trends observed by Data Center Knowledge (DCK), a leading online source of data center news and analysis, are listed below.[5] All trends, even the ones that are happening due to business reasons, are leading toward more efficient data centers.

1. Cloud Computing creates more centralized, efficient, and high-power density data centers, eliminating the distributed and inefficient smaller data centers.

---

[4] Internet World Stats, www.internetworldstats.com/.
[5] Rich Miller, Top ten data center trends of 2011, 2011, Data Center Knowledge, www.datacenterknowledge.com/top-10-data-center-trends-of-2011/.

2. A containerized modular approach is gaining popularity. This approach is more energy efficient and fundamentally different from conventional data centers.
3. Mergers between telecommunication and cloud data services companies.
4. Data centers relocating to remote locations from cities and business districts to locations that are more cost-friendly in regard to climate, tax, and electrical power.
5. Openness and sharing of information of each other's data centers is becoming prevalent. Industry has realized that sharing information is more beneficial than secretive attitudes.
6. Economizing operations and reducing energy cost is the norm.
7. Collation and wholesale operations intermingling, leading to more efficient use of space and resources.
8. Custom servers developed to specific needs making them more efficient in the process.
9. Data center infrastructure management practices leading to more efficient operations.
10. Being located in sites with renewable energy and climates that reduce energy consumption.

POWER USAGE EFFECTIVENESS (PUE)

Sustainability and energy efficiency have become important to data centers, because their use is constantly growing. The general public has become aware of the impact of data centers. On August 10, 2009, the popular comic strip *Dilbert* addressed the power and cooling of data centers (http://dilbert.com/strips/comic/2009-08-09/). The projected growth of data centers is expected to be higher than the general growth of the economy. Until 2006 there was no universal measurement of the energy efficiency of data centers. In 2006, the "Conference on Enterprise Servers and Data Centers: Opportunities for Energy Savings" was the first conference that brought focus on energy efficiency, which lead to benchmarking or measuring efficiency of data centers. General consensus was arrived at PUE. Power Usage Effectiveness (PUE) is the ratio of total power usage to the power used for computing. This measurement is simple, but extremely effective. PUE does not involve complex calculations or measurements and lengthy procedures. PUE is easy to understand and has become the data center equivalent of the nutritional facts label on food products.

$$PUE = \text{Total Facility Power} / \text{IT Equipment Power}$$

Total facility power includes:

Power consumed for computing (servers or mainframes)

Power consumed by distribution equipment such as transformers, UPS, invertors, and battery packs

Power consumed by cooling equipment such as chillers, pumps, cooling towers, and air distribution equipment

IT equipment power includes:

Power consumed by computers

IT equipment power is generally not measured with meters and cannot be determined unless separate metering equipment is installed on electrical lines, at the point just before they feed the computers. The total equipment power is generally available from the utility metering equipment, which is used to calculate the electric bills. PUE is an important matrix because establishing this measurement for both new and existing data centers presents an opportunity to improve their efficiency.

Conventional data center cooling systems have become a limitation impediment to what servers can do in the space. The need to build new data centers is not due to the increase in computing intensity, but rather to the limitations of cooling capacity. The engineering community has an opportunity to figure out an efficient way to cool these buildings, and thereby eliminate the need to construct new buildings—a contribution to sustainability in itself. Servers could perform more computing in efficiently

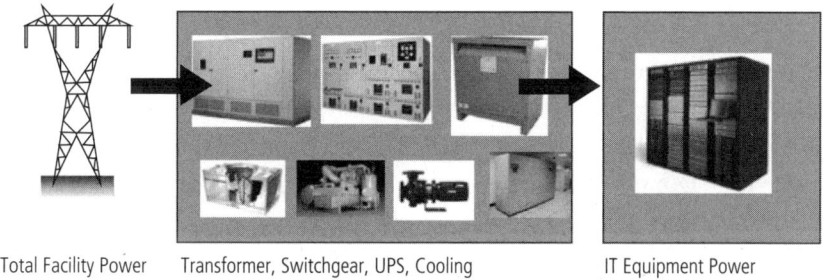

Figure 12–2 Power usage effectiveness. *Asif Syed*

Total Facility Power | Transformer, Switchgear, UPS, Cooling Tower, Chiller, and Pumps | IT Equipment Power

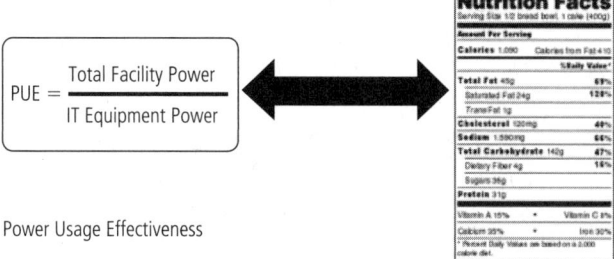

$$PUE = \frac{\text{Total Facility Power}}{\text{IT Equipment Power}}$$

Power Usage Effectiveness

cooled racks than they can when cooled by standard conventional systems. This has led to newer technologies that challenge the conventional approach.

WHY DATA CENTERS HAVE HIGH PUE OR LOW ENERGY EFFICIENCY

Cooling: Data centers are generally cooled with CRAC (computer room air conditioner) units. These are floor-standing units in the data center that supply air to the plenum below the floor. This arrangement does not take advantage of cold outside air in winter and colder months to directly cool the computers. Even with so-called free cooling CRAC units—used so as to satisfy the code mandate—the effective use of free cold from outside to cool the data center is much less than its potential.

Inefficient organization of servers: In many data centers, server racks are located haphazardly, which makes the cooling system inefficient. Air does not get effectively heated to its entire cycle. This increases the air flow, increasing the fan energy consumption. Fans are not very efficient mechanical components; they have an efficiency of approximately 65 percent. Thirty-five percent of the energy going into the fan is converted back into heat. The more air flow, the more fans, with more energy consumption and more heat gain from the fans, which needs to be cooled. An efficient arrangement of racks that creates a hot and cold aisle greatly helps reduce this inefficiency.

Hot and cold aisles not separated: Even when hot and cold aisles are created, if they are not separated, then most of the air bypasses the servers. This reduces the effectiveness of air heat gain, leading to more air flow and fan energy and addition of heat as explained in the preceding point.

Cooling people instead of electronic equipment: Usually data centers are cooled to human comfort. This practice comes from the legacy of computing equipment from the 1970s, predating the data center era. The conditions for an environment suitable for those large computing machines were similar to those necessary for human comfort. Present-day computer rooms and data centers can be maintained at much higher temperatures.

Air conditioning (AC) systems: Cooling systems such as chillers, pumps, and cooling towers have their origins in buildings other than data centers, so they were not designed with data centers in mind. Some of the features that could benefit data centers have been ignored or not used to full advantage. Economizer AC, a method of using the outdoor environment (cold) to cool interiors, can benefit data centers. When the economizer system became a code mandate, engineers had to design it, but it was considered a burden and shoehorned into buildings to comply with regulations. The economizer system's potential was never realized in buildings, and data centers have the same, or similar, inefficient systems.

Cumulative effects: The cumulative effect of these inefficiencies drastically reduces energy efficiency even further. The effect of haphazard organization of servers and of air bypassing the servers multiplies itself. The CRAC unit, which is meant for cooling the data center, is producing heat in the data center from fan inefficiency. These problems, coupled with the failure to use outside weather to its full potential, are the main reasons that data centers are so inefficient.

ELECTRICAL DISTRIBUTION

AC to DC conversion: In data centers alternating current (AC) is converted into DC, converted back into AC, and then converted to DC again in servers. This is a threefold conversion. Every time AC is converted to DC and vice versa, power is lost, dissipated as heat. This threefold conversion is done for several reasons.

1. The first conversion of AC from the utility to DC is done to store electricity in the batteries.
2. The batteries add the reliability of uninterrupted power to the servers.
3. The second conversion is from the batteries' DC to AC. This is done because servers have an AC plug connection.
4. Finally, the servers have an internal component that converts AC to low-voltage DC, to be used by the central processing unit.

In all stages of conversion, there are losses associated with rectifiers, transformers, and invertors. This is another example of the way that the fast boom of data center growth during the 1990s continues to affect the energy efficiency of data centers. During the dot-com bubble, engineers and server manufacturers did not have sufficient time to streamline the process and make it efficient.

Transformer losses: There are several transformers for voltage regulation. Starting from the utility power supply, which usually has a high voltage of 4 to 37 kilovolts, it is reduced to 480 volts. Then it is further reduced to 120 volts for servers. Servers are dual-voltage; they can use either 220 or 120 volts, like laptops. The conventional practice is to use 120 volts for servers, which creates another loss for voltage regulation. There are also isolation transformers in uninterrupted power supply (UPS) units, and losses are associated with them as well. Also, the transformer losses of computer loads (nonlinear) are almost twice the losses published for the transformers. The published losses are for mechanical loads (linear).[6]

---

[6] Philip J. A. Ling, "Overcoming Transformer Losses," Electrical Construction and Maintenance, digital edition, August 29, 2003.

TECHNOLOGIES THAT CAN BENEFIT DATA CENTER EFFICIENCY

AIR-SIDE ECONOMIZER OR FREE COOLING

Data centers use conventional CRAC units. These units use water as the cooling and heat rejection medium. Compressor-energy-free cold water can be produced from outdoors when the outdoor temperatures are low. CRAC units generally use 45°F water; to produce 45°F water, the outdoor temperature has to be about 32°F. There are relatively few hours in the year at most locations when this (32°F or lower) exists, limiting the hours of compressor-free cooling. This problem can be overcome by using air as the heat rejection medium. An air-side economizer system provides a great opportunity to extend the hours in which outside weather can be used to cool data centers. To produce air at 55°F, the outside weather has to be only 55°F, or lower by a few degrees to compensate for heat loss.

Examples of weather data for New Jersey, in the Northeast United States, are shown in Table 12–1. In the yearly total of 8,760 hours, there are only 1,527 hours below 32.5°F, whereas there are 4,427 hours below 52.5°F. A water-side economizer using 45°F water has only 1,527 hours of free cooling, whereas an air-side economizer has 4,427 hours. An air-side economizer is three times larger than a water-side economizer.

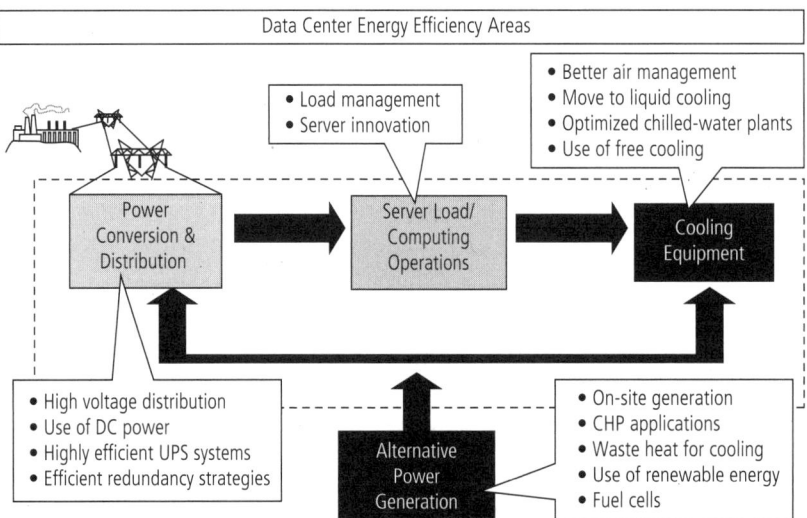

Figure 12–3 Data center energy efficiency areas, from Fact Sheet on National Data Center Energy Efficiency Information Program.   *U.S. Department of Energy (DOE) and U.S. Environmental Protection Agency (EPA), 2008*

TABLE 12–1 DRY BULB TEMPERATURE BIN FOR NEW JERSEY

Midpoint temperature in °F	Dry Bulb in °F	Total Hrs
97.5	95 to 100	6
92.5	90 to 95	40
87.5	85 to 90	122
82.5	80 to 85	500
77.5	75 to 80	620
72.5	70 to 75	847
67.5	65 to 70	671
62.5	60 to 65	927
57.5	55 to 60	600
52.5	50 to 55	730
47.5	45 to 50	634
42.5	40 to 45	513
37.5	35 to 40	1,023
32.5	30 to 35	734
27.5	25 to 30	391
22.5	20 to 25	195
17.5	15 to 20	125
12.5	10 to 15	47
7.5	5 to 10	34
2.5	0 to 5	1

LIQUID COOLING OR RACK COOLING

The rack density or power consumed in a rack has been continuously increasing. Densities of 30 kilowatts per rack are projected.[7] The density was only 4 kilowatts per rack in 2005. High densities pose a problem for conventional cooling with air delivery.

---

[7] Ron Hughes, "Data Centers of the Future," California Data Center Design Group, *Data Center Journal Online*, May 2005.

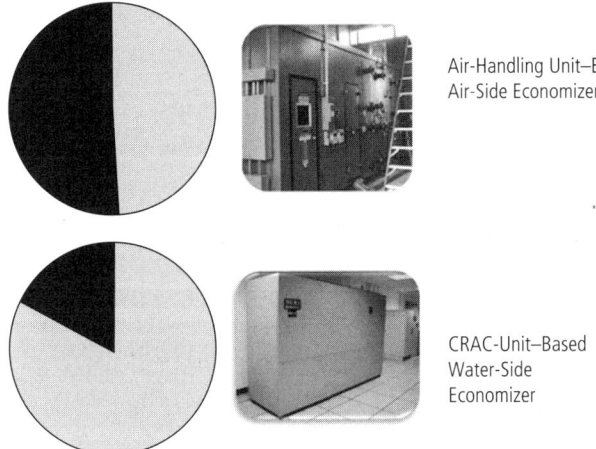

Figure 12–4 Air-side versus water-side economizer. *Asif Syed*

For New Jersey Location: Air-Side vs. Water-Side Economizer

The quantity of air required is too large to be delivered to the space. This is a good example of the way cooling has become the limiting factor in computer rooms. For example: The air quantity required for cooling 30 kilowatts is 4,500 cfm; assuming that the aisle between the racks is 3 feet, the grille size required is 9 square feet. The plenum depth needed to support these air quantities can be as high as 6 feet, depending on cable and water pipe management. For these high densities, cooling the servers at the rack makes rack cooling energy-efficient. Rack cooling is done by providing a cooling coil behind the rack, on a hinged door that opens to reach the back of the servers. The cooling coil is supplied with chilled water, which is supplied at a temperature above the dew point, to avoid condensation. Rack cooling makes the rack energy-neutral—no heat is added to the room from the rack. The energy savings come from not having to circulate large quantities of air from mechanical equipment to the racks, which saves a large amount of fan energy. The one drawback of rack cooling is that it introduces water close to the servers. But water was always used in data centers cooled by conventional methods, and in those systems it poses similar risks in case of leaks. When selecting a rack cooling or air-side economizer system, the benefits of extended hours of free cooling and the saving of fan energy must be evaluated.

In some data centers, a combination of rack cooling and air-side economizer is a good solution, as it brings the benefits of both technologies. It makes the data center flexible. The extremely-high-density racks can be rack cooled, without the limitations of air systems. The low-density racks can be air cooled, taking advantage of the extended hours of the outside air economizer.

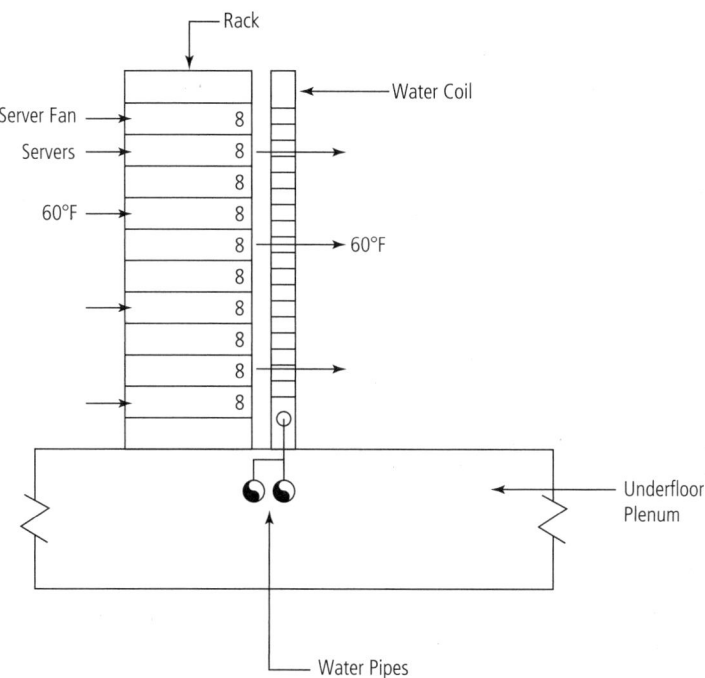

Figure 12–5 Rack cooling coil. *Asif Syed*

Rack cooling offers an opportunity to increase the capacity of existing data centers to accommodate high-density racks. It avoids capital costs for infrastructure of new building space and MEP systems to build new data centers. At a relatively low cost, just for infrastructure upgrade with chillers and refrigerant or water piping, rack or liquid cooling can be accommodated.

CARBON DIOXIDE COOLING

Cooling with carbon dioxide is similar to rack cooling. In lieu of water, carbon dioxide is used as the coolant. Carbon dioxide liquefies at high pressures and condenses at a lower temperature. In these systems, carbon dioxide is the secondary coolant, and the heat is rejected to the outdoors via a standard chiller. Liquefied carbon dioxide is circulated through the coils in the racks. Carbon dioxide evaporates in the rack coil, cooling the rack and making it heat-load-neutral to the data center. The phase change from liquid to gas in carbon dioxide cooling has a very high heat value. The heat-carrying capacity of carbon dioxide is seven times that of water.[8] This makes

---

[8] Guy Hutchins, "CO_2 Mission Critical Cooling at Imperial college, London,", TROX Advanced IT Cooling Systems.

the pipes for carbon dioxide smaller than water pipes. The main advantage of carbon dioxide rack cooling is that it eliminates water from the data center. The fear of water in the data center is strong in some sections of the data center industry. Carbon dioxide offers a good solution because it is benign to electrical components and servers. If there is a leak, it is in gaseous form. The health and safety issue of carbon dioxide leaking into the data center is addressed by installing an automatic monitoring system with alarms.

OFFICE BUILDING APPLICATIONS

Small computing spaces can be found in almost all office buildings. These can include rooms from 100 square feet to 500 square feet with servers. Most office buildings have server closets and server rooms, and some have localized data centers or even mid-tier data centers. These rooms generate heat all year-round and usually 24/7. The servers in these spaces are active even after office hours due to remote access from homes, e-mail delivery to mobile phones, backup activity, and the like. The heat generated from these spaces is cooled in summer with refrigeration. In both summer and winter, heat is rejected to the outdoors. Winter office buildings can use heat from the data center or IT rooms to heat the perimeter of the buildings, which is cold due to low outside temperatures. These data center spaces are classified[9] in Table 12–2. It is not uncommon to find localized data centers in office buildings up to 1,000 square feet. The heat from the data center can be captured and re-used both in summer and winter for domestic hot water and space heating. This permits increasing the whole building efficiency.

TABLE 12-2 DATA CENTER SPACES CLASSIFICATION

Space Type	Size	IT Equipment
Server closet	Up to 200 sq. ft.	1 to 2 servers
Server room	Up to 500 sq. ft.	2 to 15 servers
Localized data center	Up to 1,000 sq. ft.	15 to 100 servers
Mid-tier data center	Up to 5,000 sq. ft.	100s of servers
Enterprise data center	Greater than 5,000 sq. ft.	1,000s of servers

---

[9] Report to Congress on Server and Data Center Energy Efficiency, Public Law 109-431, U.S. Environmental Protection Agency ENERGY STAR Program, August 2, 2007.

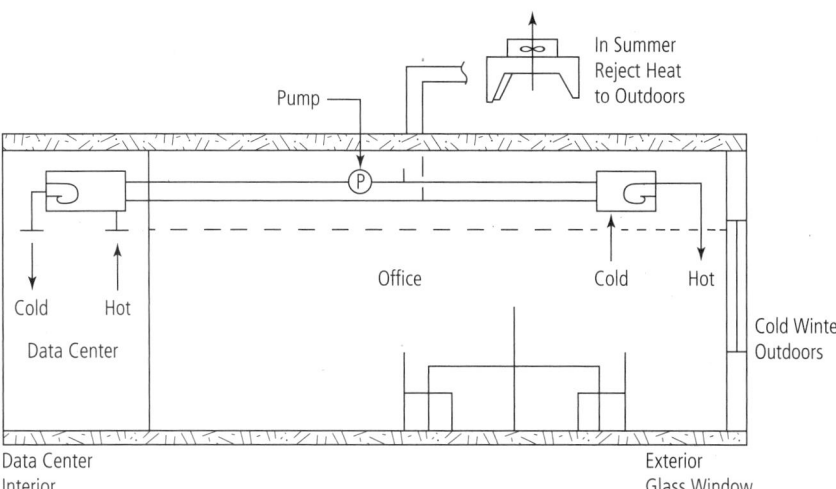

Figure 12–6 Office building data center heat transfer. *Asif Syed*

CAPTURING DATA CENTER HEAT

Data centers or computing rooms are good sources of heat energy. Almost all electricity that is fed to servers is converted into heat. This heat can be transferred to other areas in the building that require heat. The quality of heat from the data centers is low-grade heat at about 75°F. This heat can be captured or transferred to other areas via heat pumps, as in a geothermal heat system. The steady 55°F ground temperature is used in geothermal systems. The data center heat is steady for 24/7 in all seasons, like geothermal energy. Heat pumps can be used to extract the heat from the data center, and it can then be used for a variety of purposes. In cold climates, such heat can be used for heating the building and for domestic hot water heating. Although this has not been common in the data center industry, there are several recent examples. In Helsinki, heat generated from the Academica Oy data center, located in bedrock beneath Uspenski Cathedral, is used for heating homes via a district heating loop.[10] Notre Dame University's containerized data center is located adjacent to, and provides heat to, a historical greenhouse and conservatory[11] in South Bend, Indiana, where the temperature can drop to 10 to 15°F in winter.

---

[10] Tarmo Virki, "Cloud computing goes green underground in Finland," Reuters, November 29, 2009.
[11] Paul R. Brenner et al., "Grid Heating Clusters: Transforming Cooling Constraints into Thermal Benefits," Notre Dame University white paper case study, in *The Path Forward v4.0: Revolutionizing Data Center Efficiency*, proceedings of the Uptime Institute Green Enterprise IT Symposium, April 13–16, 2009, New York City.

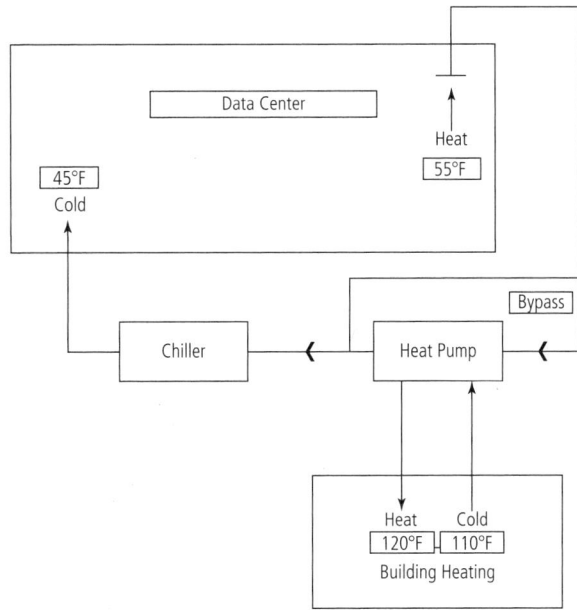

Figure 12–7 Capturing heat from a data center. *Asif Syed*

ON-SITE GENERATION AND COMBINED HEAT AND POWER

Combined heat and power is also referred to as cogeneration, or cogen. The process of cogen is more energy efficient than conventional purchase of power from the utility company and generation of heat at the site. Cogen recovers waste heat from the generation of electricity by the combustion of fuel. Utility power plants normally do not recover this heat, which is rejected into the atmosphere. Electrical load in data centers is constant and steady most of the time. The electrical power produced by cogen is used for computing electrical power in servers. The heat recovered from the cogen process on-site can be used for providing cooling to the servers. Most data centers have backup power generation in case of power failures. The equipment already there for standby generators can be put to use as cogen machines. Most of the time the standby generators are idling, as power is reliable in most developed countries. In most data centers, the only operation of standby generators is their periodic operation to exercise and prepare for their readiness in an event. This underutilized asset can be used for highly efficient, sustainable, and cost-lowering cogeneration. The cogen equipment is generally of higher quality than standby equipment. There is additional cost, above the standby equipment cost, but the ROI or payback period is very short. Cogen reduces the operating costs, increases reliability, and is environmentally friendly. Power generation on-site reduces power transmission losses from the remote power plant. The payback for cogen is very fast, generally in the range of four to seven years, depending on the type of fuel and type of technology. The cogen concept is more suitable for larger enterprise data centers than for small data centers. The benefits of cogen in data centers are:

1. Cogen yields higher energy efficiency or PUE.
2. Cogen reduces greenhouse gases and carbon footprint of the data center.
3. Steady 24/7/365 load of servers maximizes the cogen efficiency and utilization.
4. Waste heat can be used to cool the servers 24/7 through absorption air conditioning.
5. Waste heat can be used in winter to heat the buildings.
6. Standby equipment can be used for cogen.
7. Continuous operation increases reliability during startup.
8. Cogen reduces the demand on the utility grid.

AIR MANAGEMENT IN THE DATA CENTER

The biggest problem when it comes to data center air flow efficiency is lack of proper air flow. The air bypasses the servers it is meant to cool; there is insufficient cool air where it is needed, and excess cool air where it is not required. Some basic principles of air management in data centers can bring about big gains in efficiency. Generally, these principles have been overlooked because data center power and cooling densities were small. With the current pressures for energy efficiency and higher-density cooling, every Btu needs to be maximized from the air. These strategies are simple and require little technical know-how or knowledge. However, they are very effective. Some of these strategies are discussed in the following sections.

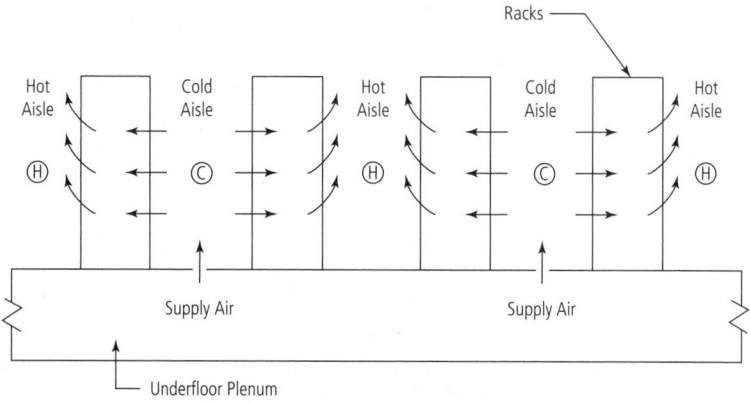

Figure 12–8 Creation of hot and cold aisle. *Asif Syed*

CREATION OF HOT AND COLD AISLES

Organizing the data center rack layout to create hot and cold aisles increases the effectiveness of cooling, thereby increasing efficiency. Most servers are located in racks that have air intake in the front and discharge in the back. If the racks are arranged with exhaust sides facing exhaust sides and intake sides facing intake sides, hot aisles and cold aisles are created. This permits air to go through its cycle of heat gain without interruption or mixing cool air with warmer air. Mixing hot and cold air in the same aisle leads to low temperature and uneven temperature spread in the data center. Some spots may be cooler than optimal, others hotter than optimal. The thermostat may be trying to cool the hot spot, thereby overcooling the entire data center. The goal is to maximize the temperature difference between the supply and return air to the servers. In a perfect situation this can be in the range of 20°F to 30°F. However, due to mixing or short cycling, most data centers see a much lower differential. This makes the system inefficient, requiring more air and more fan energy to circulate the air. The goal of the hot and cold isles is to maximize the supply and return air temperature differential.

BLANKING PANELS IN SERVERS

Server racks are generally filled from top to bottom with servers. This is an efficient arrangement, when racks are fully populated. However, some racks are not filled entirely, leading to blank spaces in the rack. These blank spaces create a bypass of cold air. The cold air mixes with the hot air and reduces the effectiveness of the air's cycle of heat gain. A simple solution to this is using blanking panels, or blank-off plates. Server rack manufacturers use these when the entire rack is not filled, to cover spaces where a unit has been removed or left out. Maintaining the design intent is a housekeeping exercise, which when done diligently can make the data center more efficient.

SEPARATION OF HOT AND COLD AISLES

The greatest loss of air-cooling capacity is from the bypassing of air. After the server bypass, the isle bypass is another inefficiency in the data centers. Air is cooled in the air conditioning units so it can cool the servers. When a portion of air entirely bypasses the rack from the cold side to the warm side, it greatly reduces the cooling efficiency. This leads to the use of additional fan power to circulate additional air. The compressors also operate inefficiently, as they do not see the design temperature difference. Separation between hot aisle and cold aisle, or containment of either hot aisle or cold aisle, can lead to a reduction in air flow. At 25 percent air flow reduction the fan power reduction is 60 percent.[12]

---

[12] Mukesh Khattar, "Energy Efficiency in Data Centers," Climate Leaders Meeting, Boulder, CO, December 2007, Oracle. www.epa.gov/climateleaders/documents/events/dec2007/Mukesh_Khattar.pdf.

INDEX

A

Acoustics
   importance, 53–54
   theater usage, 57
Active chilled beams, 68–70
   coverage, 77t
   example, 70f
   nozzle, induction effect, 70f
   passive chilled beams, contrast, 73–74
Air circulation, impact, 65
Air cleanliness, method, 8
Air distribution, problem, 91f
   correction, 92f
Air inefficiency, 64
Air reheat loads, impact, 81
Air-side economizer, 230
Airside economizer, water side
      economizer (contrast), 232f
Alternating current (AC) to direct
      current (DC) conversion, 229
American Institute of Architects (AIA),
      hospital guidelines, 51–52
American National Standards Institute
      (ANSI), energy code 90.1,
      118–128
   glazing performance requirements,
      123t
American Society of Heating
      Refrigeration and Air
      Conditioning Engineers
      (ASHRAE), energy code 90.1, 13,
      118–128
   glazing performance requirements,
      123t
Annual solar electric power generation,
      173–174

Aquifers
   energy storage, 148f
   usage, 147

B

Bi-fuel engines, conversion kits, 214
Boiler energy savings, 64
Building management and security
      (BMS) systems, 129–130
Buildings
   average site energy consumption, 176t
   construction, importance, 2
   cooling, night cool (usage), 151–153
   core spaces, UFAD (impact), 96
   energy consumption, 174–176
      impact, 1
      percentages, 2f
   energy efficiency, improvement,
      181–183
   energy end-use splits, 15–16
   energy performance (enhancement),
      intermediate cavity (usage), 134
   envelope criteria, codes, 119–120
   external heat load, reduction, 73
   floor-to-floor height, impact, 95–96
   heat dissipation, aquifers (usage), 147
   improvements, opportunities, 4–5
   operable windows, 113
   systems, efficiency level, 16
   technologies, potential, 5
   thermal energy storage, 145
   UFAD, impact, 95–100

C

Capillary tubes, example, 38f
Carbon dioxide
   cooling, 233–234
   heat pumps, 202–203
Carbon Disclosure Rating, 17
Carbon footprint, 17
Cast-iron radiators, DIU retrofit
      (example), 110f
Cattle farms, 207–208
Ceilings
   chilled beams, usage, 74
   diffusers, ventilation effectiveness,
      88t, 108t
   space, chilled beams (impact), 66
Chilled beams
   applications, 72–77
   benefits, 63–67, 79
   ceilings, impact, 74
   comfort/noise, 65
   design, to-do list, 73–75
   economics, 74
   energy savings, 63–65
   example, 62f
   geothermal systems, coupling, 64–65
   humid climates, impact, 74–75
   installation, example, 75f
   layout, 73
   maintenance level, 66–67
   moisture generation, 77
   operation/technology, principle, 62–63
   performance, 81–82
   space savings, 65–66
   system, flexibility, 66
   types, 67–72
   usage, 61–62
      underfloor air distribution (UFAD)
      applications, combination, 78–82
   wattage, determination, 76

Chilled walls/ceilings
    capillary systems, examples, 38f
    usage, 25
Chiller energy savings, 64
Churn, chilled beams (impact), 66
Classrooms
    design, acoustics (importance), 53–54
    displacement/ceiling diffusers,
        ventilation effectiveness, 108t
    displacement ventilation
        examples, 54f–55f
        usage, 53–56
    DIUs
        example, 111f
        usage, 108–111
    induction units, examples, 110f
    speech intelligibility, importance, 54
Climate zones, 119–121
    example, 120f
Closed-loop systems, types, 193t
Cloud computing, 225
Cogeneration, 205
    applications, 207–211
    benefits, 208–211, 237
    capacity range, 212t
    DSG/heat recovery, combination, 215
    economics, 206–207
    efficiency, increase, 208
    electrical power/total efficiencies, 212t
    feasibility study, 222
    operating cost, reduction, 208–210
    operational efficiency, 209f
    plant
        examples, 218f
        location, 206
    power plant efficiency, 207f
    technologies, 211–221
Cold aisles
    creation, 237f, 238
    hot aisles, separation, 238
College dormitories, radiant ceiling
    panels (usage), 36
Combined heat and power (CHP)
    generation, 205
    on-site generation, 236–237

COMcheck, 122–123
Commercial heat pump, 199f
Commercial offices, chilled beams
    (usage), 75–77
Computational fluid dynamics (CFD)
    analysis, 91
Concrete flat slab, floor-to-floor height
    savings, 96f
Concrete structure buildings,
    floor-to-floor height (impact),
    95
Continuous generators, 215–216
Conventional air conditioning systems,
    153–156
    diurnal thermal storage, usage
        (example), 153f
Conventional air distribution systems
    displacement systems
        benefits, 46t–47t
        contrast, 42–47
    floor-to-floor height, example, 95f
    UFAD systems, contrast, 90t
Conventional overhead air distribution,
    schematic, 84f
Conventional power plant efficiency,
    206f
Cooling towers, usage, 155–156
    example, 156f

D
Data centers
    air management, 237–238
    cogeneration, benefits, 237
    efficiency
        areas, 230f
        technologies, impact, 230–234
    EPA study, 224–225
    heat capture, 235, 236f
    heat transfer, 235f
    history, 224–225
    low energy efficiency, 228–229
    PUE levels, 228–229
    spaces, classification, 234t
    sustainability, 223
    trends, 225–226

Daylight harvesting system, integration,
    133
Daytime solar radiation, night cool
    (contrast), 150–151
Developing countries, growth
    (acceleration), 4
Direct digital control (DDC) systems,
    38–39
Direct solar gain, reduction, 130
Dispatchable standby generation (DSG),
    214–215
    benefits, 214t
Displacement
    diffusers, ventilation effectiveness,
        88t, 108t
    distribution, underfloor air
        distribution (contrast), 47–48
    principle, 43f
    systems, UFAD systems (contrast),
        48f
    temperature gradient, 43f
Displacement induction
    contaminant control, 106f
    duct/piping arrangement, 109f
    principle, example, 102f
    units, induction units (contrast), 107f
Displacement induction units (DIUs),
    101
    air requirement, 102
    applications, 107–113
    benefits, 103–106
    electrical costs, reduction, 105
    energy consumption, 103–104
    examples, 109f–111f
    fan coil system, retrofit (example),
        110f
    indoor environment improvement,
        105
    maintenance, reduction, 105–106
    noise levels, 104
    space savings, 105
    thermal comfort, 104
    Trox DIU, example, 103f
Displacement ventilation
    acoustics, impact, 53, 56

applications, 48
benefits, 44–46
conventional air distribution system
    benefits, 46t–47t
    contrast, 42–47
energy savings, 44–45
examples, 49f–50f, 54f–55f, 58f
explanation, 42
history, 41–42
indoor environment, 45–46
large public spaces, 48–59
mixed-air systems, contrast, 42–47
office space usage, 59
performance space usage, 56–59
systems
    supply air temperature, 42
    underfloor air distribution (UFAD), contrast), 48t
teaching environment/classroom usage, 53–56
theater usage, 56–59
thermal stratification, 45
Diurnal energy storage, 148
Diurnal thermal storage
    cooling towers, usage (example), 156f
    UFAD systems, usage, 155f
    usage, example, 153f
Double-skin envelope, 128, 134–136
Double-wall façade
    airflows, 135f
    dampers, 135f
    engagement methods, 135–136
    heat harvesting, 138f
    solar radiation, impact, 150f
    summer/winter uses, 151f
Dry bulb temperature bin (New Jersey), 231t
Dual-fuel engines, conversion kits, 214

E

E coating, effect, 128t
Economic development, importance, 1–2
Electrical distribution, 229
Electrical rooms, UFAD usage (example), 97f
Electricity, grids, 171–172
Embodied energy
    comparison, 18f
    definition, 17
    operational energy, contrast, 17–19
    percentage, 18–19
Energy
    consumption, 174–176
        growth, technologies (impact), 2–3
        reduction, 130, 165, 183
    cost budget method, 123
    off-site export, 171–172
    peak demand, reduction, 140t
    reduction, 140t
        overhangs, usage, 131t
    resources, pressure, 4
    storage, 171
    use, reduction, 130–131
Energy codes
    envelope compliance, 121–122
    options, 122t
    U.S. adoption, 119f
Energy efficiency
    funding opportunities, 19–20
    improvement, 181–183
    increase, 164–165
    profits/savings, 11–12
ENERGY STAR program, 183
Engaging envelopes, 116
Envelopes
    code performance, exceeding, 128–143
    criteria, code mandates, 119–120
    energy code compliance, 121–122
    glazing characteristics, 123–128
    light to solar gain (LSG) ratio, 127–128
    performance, improvement, 123–124
    simulation, energy cost budget method, 123
    solar heat gain coefficient (SHGC), 126
    visible light transmittance (VLT), 126
Evacuated tube collectors, 168–169
    example, 169f
External shades, 128, 130–132
    examples, 132f

F

Façade cooling load, percentage, 118f
Fans
    equipment efficiency, radiant cooling factor, 24
    inefficiency, 64
Fins, 128
Flat plate collectors, 168
    example, 169f
Floor-mounted console heat pump, 200f
Floor-to-floor height
    impact, 95–96
    savings, 96f
Fossil fuel resources, 8–10
Frame U values, reduction methods, 125
Free cooling, 230
Free heat, availability, 191
Fuel cells
    cogeneration, 219–221
    operation, 220f
    systems, comparative cost, 221t
    usage, 212
Fume hood exhaust, impact, 82

G

Gas turbines
    applications, 217–218
    cogeneration, 216–219
        example, 217f
    usage, 211
Geothermal energy, heat production, 187–189
Geothermal heat pumps, 190–198
    efficiency
        reasons, 190–191
        sources, 190–191
    energy efficiency, 192f
    operation
        cooling mode, example, 188f, 190f
        heating mode, example, 189f, 191f

Geothermal heat systems, HVAC equipment (comparison), 147–148
Geothermal methods, 191–193
Geothermal resource map, 186f
Geothermal systems, 185
   building benefits, 187
   chilled beams, coupling, 64–65
   classification, 192
   economic benefits, 187
   environmental benefits, 186–187
Glare (reduction), overhangs (usage), 131t
Glazing assembly u-value, 124–125
Glazing characteristics, 123–128
   list, 124t
Glazing u-value, 124–125
Greenhouse gases (GHGs), 10–11
   effect, 10f
   emitting countries, aggregate contributions, 11f
   identification, 10
Greenhouses, 207
Green roofs, 128

H

Health-care facilities
   displacement ventilation, usage, 50–52
   DIUs, usage, 111–112
Heat carrying capacity, radiant cooling factor, 24
Heat energy, quality, 187–188
Heat extraction/sinking, 191–193
Heat harvesting, 138f
Heat-pipe evacuated collector, 169f
Heat pumps, types, 198–203
Heat transfer, 201f
Heat transmission coefficient, 124–125
High-grade energy, usage, 188
High-performance envelope, 115
   definition, 117–118
High-temperature solar thermal collectors, 168
Horizontal closed-loop heat pump, 195f
Horizontal closed-loop system, 194

Hospitals
   American Institute of Architects (AIA) guidelines, 51–52
   chilled beams
      benefits, 79
      UFAD, combination usage, 79–80
      usage, example, 80f
   codes, 51–52
   displacement ventilation, usage, 50–52
   energy, usage, 51f
   infection control, importance, 51
   radiant ceiling panels, 38–39
   systems, history, 52
Hot aisles
   cold aisles, separation, 238
   creation, 237f, 238
Human thermal comfort, variables, 23–24
Humidity control strategy, 73
Hydrogen fuel cells, 220
Hypocausts
   example, 22f
   usage, 21, 23

I

Indoor air
   environment, UFAD (impact), 87–88
   exhaust, 138–139
   supply, 137–138
Indoor air quality
   guarantee, 65
   improvement, strategies, 7
   UFAD, impact, 87–88
Indoor building environment, cleanliness (method), 8
Indoor environment
   displacement ventilation, usage, 45–46
   DIUs, impact, 105
Induction principle, suitability/adaptability, 69
Induction units
   displacement induction units, contrast, 107f
   history, 106–107
   usage, 102

Information technology (IT)
   power, 227
   rooms, UFAD usage (example), 97f
Insulated glass unit (IGU), 124
   SHGC values, 126t
   U values, 125t
International Standards Organization (ISO), Standard ISO 7730, 23

K

Kyoto Protocol, greenhouse gas identification, 10, 17

L

Laboratories
   chilled beams/UFAD, combination usage, 80–82
      example, 82f
   energy usage, 81
   radiant ceiling panels, usage, 34–35
   set points, 81–82
Lake closed-loop system, 195–196
Large public spaces, displacement ventilation (usage), 48–59
Las Vegas, high/low temperatures, 152t
Leadership in Energy and Environmental Design (LEED)
   rating systems energy optimization points, 14t
   Reference Guide for Green Building Design and Construction UFAD information (2009), 84
   version 3 (LEED V3), 13–14
Life-cycle cost analysis, familiarity, 6
Light glare, reduction, 130
Light to solar gain (LSG) ratio, 127–128
Liquid cooling, 231–233
Low-grade energy heat, abundance/availability, 189
Low-temperature solar thermal collectors, 168

M

Mechanical, electrical, and plumbing (MEP) systems, improvements, 4–5

INDEX 243

Medium-grade energy, limitation, 188–189
Medium-temperature solar thermal collectors, 168
Micro-cogeneration or combined heat and power (micro-CHP), 221–222
Micro-turbines
  cogeneration, 219
  usage, 212
Mixed-air systems, displacement systems (contrast), 42–47
Mullion U values, reduction methods, 125
Multidirectional radiant slabs, 37f
Multiservice chilled beams, 70–72
  benefits, 70–71
  examples, 71f–72f
  installation, example, 71f

N

National Renewable Energy Laboratory (NREL), geothermal resources maps, 186
Natural gas industrial prices, 210t
Natural gas sites, availability, 213–214
NetMetering, 179–181
Net metering, 180–181
Net-zero buildings, 163
  analysis, 178
  consumption, reduction, 183–184
  energy efficiency, improvement, 181–183
  energy simulation, 178
  feasibility, 178–179
  financial aspects, 182
  mechanical/electrical systems, 181–183
  process, steps, 164, 166, 181, 183
  solar energy, 177–181
    impact, 165f
  technologies, impact, 182–183
Net-zero definition, 179–181
New Jersey, dry bulb temperature bin, 231t

New York City, high/low temperatures, 152t
Night cool, 150–153
  UFAD, impact, 154–155
  usage, examples, 151–153
Noise levels, 65
  DIUs, impact, 104
Nonengaging envelopes, 116
Nonrenewable energy, partial thermal storage system, 158f
Nonrenewable energy storage, 156–161
Nozzle, induction effect, 70f

O

Office buildings
  active chilled beam coverage, 77t
  data centers
    applications, 234–237
    heat transfer, 235f
  energy reduction, 5
  high load density, 77t
  office-sensible (non moisture) space heat loads, 77t
  radiant ceiling panels, 31–33
    examples, 32f
    finishes, 33f
Office layout (configuration), UFAD (impact), 86–87
Office spaces, displacement ventilation (usage), 59
Off-site energy export, 171–172
Once-through system, 82
On-side generation and combined heat and power, 236–237
Open-loop geothermal wells, usage (factors), 197–198
Open-loop systems
  types, 193t
  water quantities, 197t
Open-well open-loop system, 196–198
Operable windows, 113
  controls, 129f
  sensors, 129f
  usage, 128–130

Operational energy
  comparison, 18f
  definition, 17
  embodied energy, contrast, 17–19
Outdoor air
  buffer, 136
  supply, 136–137
Outdoor chilled water storage tank, example, 160f
Overhangs, 128, 130–132
  usage, 131t

P

Partial storage system operation, 159t
Partial thermal storage system, 158f
Passive chilled beams, 67–68
  active chilled beams, contrast, 73–74
  benefits, 68
  schematic, 68f
  UFAD application, 78f
Patient rooms
  chilled beams, benefits, 79
  chilled beams/UFAD, combination usage, 79–80
  DIUs, usage, 111–112
Peak cooling load, reduction, 140–142
  illustration, 141f
Per capita energy consumption, 14–15
  world BTU consumption, 15f
Performance spaces, displacement ventilation
  example, 58f
  usage, 56–59f
Perimeter buildings
  DIUs, example, 113f
  DIUs, usage, 112–113
Perimeter heating system, 74
Perimeter zone heat pumps, usage, 201
Photovoltaic power generation, 174t
Photovoltaic system, efficiency, 170
Plenum
  design, 56
  integrity, 96, 99
Polyethylene (PEX) tubing, 25, 25f
Pond closed-loop system, 195–196

Power usage effectiveness (PUE), 226–229
  example, 227f
Prime generators, 215–216
Process loads, impact, 81
Public spaces
  displacement ventilation, examples, 49f–50f
  radiant cooling floors, 28–30
    benefits, 28–29
PVWatts software, printout, 175t

Q
Quality of life, benefits, 7–8

R
Rack cooling, 231–233
  coil, 233f
Radiant ceiling panels, 31–33
  applications, 34–39
  construction, 33–39
  examples, 32f
  finishes, 33f
Radiant cooling
  applications, 28–33
  benefits, 26
  condensation, impact, 26
  conventional systems, differences, 27t
  example, 22f
  explanation, 23–26
  historic preservation, impact, 39
  history, 21–23
  systems
    benefits, 32
    energy savings, 27
  technologies, 24–25
  usage, reasons, 26–27
Radiant cooling floors, 28–30
  benefits, 28–29
  construction, 30–31
  examples, 29f–30f
  floor sandwich, 31f
  walkway bridge, example, 29f
Radiant floors, usage, 24–25

Radiant panels, usage, 24
Radiant pipe manifold, example, 30f
Raised floor, air distribution problem, 91f
  correction, 92f
Reciprocating engines
  cogeneration, 212–216
    example, 216f
Reciprocating engines, usage, 211
Renewable energy storage, 146–153
REScheck, 122–123
Residential buildings, radiant ceiling panels (usage), 35–36
Rooftop heat pump, 200f
Room air distribution, induction principle (suitability/adaptability), 69

S
Seasonal energy storage, 146–148
  schematic, 147f
Servers, blanking panels, 238
Sewage treatment plants, 207
Sick building syndrome (SBS), causes, 7
Site energy, 174–175
Site-to-source conversion table, 13t
Site-to-source effect, 12–13
Slinky closed-loop system, 194–195
  ground installation, 195f
Slinky loop, pier installation, 196f
Slinky loop system, lake installation, 196f
Solar electric power, annual generation, 173–174
Solar energy, 163
  abundance, 164
  harvesting, 166
  impact, 165f
Solar heat gain coefficient (SGHC), 126
  values, 126t
Solar intensity, 172–173
  U.S. locations, 173t
Solar photovoltaic cell, efficiency, 166–167

Solar photovoltaic energy efficiency, 167f
Solar photovoltaic system, solar thermal system (contrast), 170–172
  list, 171t
Solar radiant system, integration, 133
Solar radiation, 149–150
  impact, 150f
Solar-responsive internal blinds/shades, 128, 133–134
  position, 133
Solar-responsive shading system, inputs/outputs, 134t
Solar spectrum, low-E (relationship), 127f
Solar thermal collectors, efficiency, 169
Solar thermal energy
  efficiency, 170f
  harvesting, 167–169
Solar thermal panels, types, 168
Solid State Energy Conversion Alliance (SECA), 220
Source-to-site energy conversion factor, 176–177
  EPA national average, 177t
Space, ventilation, 42
Standby generators, 215–216
Standby power, facilities usage, 215
Steam turbines, usage, 212
Steel structure buildings, floor-to-floor height (impact), 95
Stored thermal energy, source, 146
Sustainable energy systems, funding opportunities, 19–20
Sustainable technologies, adoption (cost), 6

T
Teaching environment
  displacement/ceiling diffusers, ventilation effectiveness, 108t
  displacement ventilation
    examples, 54f–55f
    usage, 53–56
  DIUs, usage, 108–111

Technologies
  cost, 6
  funding opportunities, 19–20
Theaters
  acoustics, 57
  air outlets, 56–57
  displacement ventilation
    example, 58f
    usage, 56–59
  lighting, 57
  plenum, design, 56
  return air, 57
  stages, 58
Thermal comfort
  DIUs, impact, 104
  ISO definition, 23
  UFAD, impact, 87
Thermal density, 142t
Thermal energy storage, 145, 157–161
Thermal mass, 128, 140
  diurnal temperature, relationship, 142f
  energy reduction, 142–143
  partition type, 141t
  reduction, 140–141
Thermal storage systems
  space requirements, 161
  types, 158
Tint, effect, 128t
Total facility power, 227
Transformer losses, 229
Triple-pane glass, 128, 139
Triple-pane windows, 139
Trombe walls
  schematic, 149f
  usage, 149–150
Trox displacement flow diffusers, 44f
Trox DIU, example, 103f

U
Underfloor air distribution (UFAD), 83
  air leakage, 96, 99
  air requirements, 94
  benefits, 85–88
  calculations, 94

chilled beams, usage (combination), 78–82
churn costs, 86–87
conventional air distribution systems, contrast, 90t
design, 96–100
  integration, 99
diffuser, example, 89f
displacement distribution, contrast, 47–48
displacement ventilation systems, contrast, 48t
diurnal thermal storage, usage, 155f
ductwork, requirements, 94
electrical/information technology room usage, example, 97f
energy efficiency, 85
examples, 89f, 98f–99f
flexibility, 86–87
floor-to-floor height, 94
  example, 95f
impact, 95–100
incremental cost, 93t
indoor air environment/quality, 87–88
LEED 2009 Reference Guide for Green Building Design and Construction information, 84
mechanical systems component design, 99–100
myths, 94
night cool, usage, 154–155
occupant control, 88
office restructuring, 86–87
plenum integrity, 96, 99
schematic, 84f
stair section, 97f
systems
  cost, 92–93
  displacement systems, contrast, 48f
  toilet section, 97f
theater usage, 56–57
thermal comfort, 88
validation, CFD analysis (usage), 91–94

Unidirectional radiant slabs, 37f
Unknown technology factor, impact, 5–6
U.S. annual solar intensity, 173t
U.S. buildings energy end-use splits, 16f
U.S. climate zones, 120f
U.S. coal resources/reserves, 1f
U.S. geothermal resource map, 186f
U.S. primary energy production, 9f
U-value, 124–125
U values, reduction methods, 125

V
Vacuum tube collectors, 168–169
  example, 169f
Variable process loads, impact, 81
Vertical closed loop-system, 193–194
Vertical floor-mounted heat pump, 199f
Visible light transmittance (VLT), 126–127

W
Waste heat, recovery, 213
Water side economizer, airside economizer (contrast), 232f
Water temperature difference/flow rate, 197t
Water-to-air heat pumps, 198–201
  types, 198t
Water-to-water heat pumps, 202
  example, 202f
Well field, layout/capacity, 194f
World energy consumption, shares, 3f
World Internet penetration rates, 225f
World marketed energy consumption, 3f
World per capita energy consumption, 15f
Worldwide economic growth, impact, 4

Z
Zero energy buildings (ZEB), 179

JOHN WILEY & SONS, INC. provides must-have content and services to architecture, design and construction customers worldwide. Wiley offers books, online products and services for professionals and students. We are proud to offer design professionals one of the largest collections of books on sustainable and green design. For other Wiley books on sustainable design, visit www.wiley.com/go/sustainabledesign

ENVIRONMENTAL BENEFITS STATEMENT

This book is printed with soy-based inks on presses with VOC levels that are lower than the standard for the printing industry. The paper, Rolland Enviro 100, is manufactured by Cascades Fine Papers Group and is made from 100 percent post-consumer, de-inked fiber, without chlorine. According to the manufacturer, the use of every ton of Rolland Enviro100 Book paper, switched from virgin paper, helps the environment in the following ways:

Mature trees	Waterborne waste not created	Water flow saved	Atmospheric emissions eliminated	Soiled Wastes reduced	Natural gas saved by using biogas
17	6.9 lbs.	10,196 gals.	2,098 lbs.	1,081 lbs.	2,478 cubic feet